Implicit
Filtering

SOFTWARE • ENVIRONMENTS • TOOLS

The SIAM series on Software, Environments, and Tools focuses on the practical implementation of computational methods and the high performance aspects of scientific computation by emphasizing in-demand software, computing environments, and tools for computing. Software technology development issues such as current status, applications and algorithms, mathematical software, software tools, languages and compilers, computing environments, and visualization are presented.

Software, Environments, and Tools

Implicit Filtering

C. T. Kelley
North Carolina State University
Raleigh, North Carolina

Society for Industrial and Applied Mathematics
Philadelphia

Figure 9.1 reprinted with permission from Elsevier.
Tables 9.1–9.3 reprinted courtesy of Katherine Fowler.

Library of Congress Cataloging-in-Publication Data

Kelley, C. T.
  Implicit filtering / C. T. Kelley.
      p. cm. – (Software, environments, and tools ; 23)
  Includes bibliographical references and index.
  ISBN 978-1-611971-89-7
  1. Industrial engineering–Mathematics. 2. Filters (Mathematics) 3. Prediction theory.
I. Title.
  T57.K45 2011
  601'.5196–dc23

2011016187

 is a registered trademark.

Contents

Preface

This book is an introduction to implicit filtering, one of many derivative-free optimization methods which have been developed over the last twenty years. The audience for this book includes students who want to learn about this technology, scientists and engineers who wish to apply the methods to their problems, and specialists who will use the ideas and the software from this book in their own research.

Implicit filtering is a hybrid of a projected quasi-Newton or Gauss–Newton algorithm for bound constrained optimization and nonlinear least squares problems and a deterministic grid-based search algorithm. The gradients for the quasi-Newton method and the Jacobians for the Gauss–Newton iteration are approximated with finite differences, and the difference increment varies as the optimization progresses. The points on the difference stencil are also used to guide a direct search.

Implicit filtering, like coordinate search, is a **sampling method**. Sampling methods control the progress of the optimization by evaluating (sampling) the objective function at feasible points. Sampling methods do not require gradient information but may, as implicit filtering does, attempt to infer gradient and even Hessian information from the sampling.

imfil.m is a MATLAB implementation of the implicit filtering method. This version differs in significant ways from our older Fortran code [29]. This document is a complete reference to version 1.0 of **imfil.m**, covering installation, testing, and its use in both serial and parallel environments.

As implicit filtering has evolved since its introduction [131], so have several related approaches. The current version of implicit filtering, as reflected in this book and in **imfil.m**, uses ideas from [7, 38, 69, 92].

The plan of the book is that Chapters 2, 6, and 8 will also serve as the stand-alone users' guide to **imfil.m**.

Chapter 3 has two functions. The first is to give students who are using the book in a course or in their research enough background to enable them to put the algorithms in perspective and to understand what some of the options for **imfil.m** do. The second purpose is to introduce the notation needed to follow the algorithmic and theoretical development in Chapters 4 and 5.

Chapter 4 is an overview of the algorithms in **imfil.m** and an explanation of some of the design decisions. This chapter also illustrates the details of many of the algorithmic parameters one sets in the `options` structure. We develop the convergence theory for **imfil.m** in Chapter 5.

Finally, in Part IV of the book we show how **imfil.m** can be applied in the context of a few problems. One of these problems is intended to be very simple (Chapter 8), and the source code should be easy for the reader to modify and play with. The other two case studies arose from research projects [28, 28, 53], and the codes for the applications were not written with this book in mind. Hence, the application code for Chapters 9 and 10 should be viewed as "black boxes." The driver codes which call **imfil.m** to solve the problems, on the other hand, should be easy to modify.

This book owes its existence to my students and collaborators who worked on the algorithm, the Fortran code [29], and the applications which drove the development. This list of implicit filtering heroes includes Astrid Battermann, Griff Bilbro, Greg Characklis, Tony Choi, Todd Coffey, Gilles Couture, Robert Darwin, Joe David, Karen Dillard, Owen Eslinger, Matthew Farthing, Dan Finkel, Katie Fowler, Joerg Gablonsky, Paul Gilmore, Deena Hannoun, Lea Jenkins, Brian Kirsch, Anna Meade, Casey Miller, Dave Mokrauer, Alton Patrick, Jill Reese, Dan Stoneking, Mike Tocci, Bob Trew, and Tom Winslow.

Sampling methods like implicit filtering have evolved significantly in the last several years, and **imfil.m** has been vastly improved by my interaction with master searchers and interpolators such as Mark Abramson, Charles Audet, Andrew Booker, Andrew Conn, John Dennis, Genetha Gray, Tammy Kolda, Michael Lewis, Jorge Moré, John Nelder, Chung-Wei Ng, Mike Powell, Luis Rios, Nick Sahinidis, Katya Schienberg, Christine Shoemaker, Virginia Torczon, Luis Vicente, Stefan Wild, and Margaret Wright.

The development of implicit filtering has been supported by several grants from the National Science Foundation and the Army Research Office, most recently NSF grants DMS-0707220, CDI-0941253, and OCI-0749320, ARO grants W911NF0910159, W911NF-07-1-0112, and W911NF-06-1-0412, and USACE contracts W912HZ-10-P-0221 and W912HZ-10-P-0256.

C. T. Kelley
Raleigh, North Carolina
March 2011

How to Get the Software

I maintain and update the software. The examples in this book were done with version 1.0. The code will evolve as I and others use it in applications, hence we do not include source code as part of the book.

You can get the latest version of the software from

http://www.siam.org/books/se23/

On that page you will find the following:

- A pdf file of the users' guide [78]. The users' guide is Chapters 2, 6, and 8 of this book.

- **imfil.m**
  This is the main implicit filtering code.

- **imfil_optset.m** handles the options.

- Several examples within the `Examples` directory:

  - simple example from § 2.4 in the `Simple_Example` subdirectory;

  - example for linear constraints from § 7.1 and § 7.4.1 in the `Linear_Constraints` subdirectory;

  - case study for parameter identification problem for the simple harmonic oscillator (see Chapter 8) in the `Case_Study_PID` subdirectory;

  - case study for the hydrology example (see Chapter 9) in the `Case_Study_HC` subdirectory;

  - case study for the water resources policy example (see Chapter 10) in the `Case_Study_Water` subdirectory.

- The `Imfil_Tools` directory has examples and useful programs for the advanced options in Chapter 7.

You can download the whole works as a .tar.gz file. If you do that, the examples are in clearly labeled subdirectories.

You can obtain MATLAB from

The MathWorks, Inc.
3 Apple Hill Drive
Natick, MA 01760
(508) 653-1415
Fax: (508) 653-2997
info@mathworks.com
http://www.mathworks.com

Part I

Preliminaries

Chapter 1

Introduction

This book is about implicit filtering, a deterministic sampling method for bound constrained optimization. Methods like implicit filtering are designed to minimize functions which are noisy, nonsmooth, possibly discontinuous, or random and which may not even be defined at all points in design space. These are not methods for smooth problems where the function to be minimized and the constraints are expressed in terms of differentiable functions.

We will discuss the algorithmic details and the convergence theory for implicit filtering and put that in the context of an implementation of the method in MAT-LAB. In this introductory chapter we describe sampling methods in very general terms and motivate their use. In Chapter 2 we present an overview of the MATLAB code **imfil.m**, describe the kinds of problems implicit filtering as implemented in **imfil.m** is designed to solve, and show how to run **imfil.m** to solve these problems.

The later chapters develop the theory behind the algorithm and explain the algorithmic decisions we made in developing **imfil.m**. There has been significant activity in sampling methods for optimization (i.e., methods which drive the optimization only by evaluating the objective function and do not evaluate gradients or Hessians), and implicit filtering is part of that activity. We draw some of the connections to other approaches in the chapters on theory and algorithmic design.

Finally, in Part IV, we present three case studies which demonstrate how one can use **imfil.m**. Chapter 8 is about a simple model problem where the code is easy to understand and modify. The applications in Chapters 9 and 10, on the other hand, are based on recent research papers [28, 53] and show how **imfil.m** can be used with "black-box" functions which have internal randomness or discontinuities or are not everywhere defined.

1.1 What Is the Problem?

Implicit filtering solves **bound constrained optimization** problems,

$$\min_{x \in \Omega} f(x), \tag{1.1}$$

by which we mean that the goal is to minimize the **objective function** f subject to the condition that $x \in R^N$ is in the **feasible region** (or **nominal design space**)

$$\Omega = \{x \in R^N \mid L_i \leq (x)_i \leq U_i\}, \tag{1.2}$$

which is a **hyperrectangle** in R^N. In (1.2) L_i and U_i are the (finite) upper and lower bounds on the ith component $(x)_i$ of the vector x. **imfil.m** uses the bounds in its internal scaling, and we require that the bounds be finite for that reason.

If the problem is a nonlinear least squares problem, i.e.,

$$f(x) = \|F(x)\|^2 / 2,$$

where $F : R^N \to R^M$ and $\| \cdot \|$ is the Euclidean norm on R^M, then implicit filtering uses methods which exploit that structure. The function F is called the least squares residual.

If f is a smooth function, the gradient-based methods we describe in Chapter 3 are effective solvers. If, however, f can be evaluated only to low accuracy or the evaluation is noisy, f has discontinuities, or f may fail to return a value, then the conventional methods will fail. This book is about one approach to these difficulties.

1.2   Sampling Methods

Implicit filtering is a **sampling method**. By this we mean that the optimization is controlled only by evaluating f at a cluster of points in Ω. That evaluation determines the next cluster. The classical coordinate search method is one example of a sampling method. **imfil.m** is a MATLAB implementation of implicit filtering.

We will draw a distinction between pure sampling methods—such as the Nelder–Mead [105] and Hooke–Jeeves [68] algorithms, coordinate search, and the many variations of these methods [7, 45, 75, 88]—and what we will call interpolatory methods, where first (and often second) derivative information is harvested from the sampling by interpolation [36, 38, 63, 112, 114]. Implicit filtering lies in the middle and seeks to exploit the advantages of pure sampling for discontinuous and highly oscillatory problems, while using first-order interpolation (via linear least squares) and a quasi-Newton (see § 3.8) or Gauss–Newton (see § 3.9.1) model Hessian to capture the rapid local convergence of interpolatory methods for problems which are well modeled by smooth approximations. Implicit filtering also differs from the approach in [102], which samples $O(N^2)$ points and computes difference Hessians.

1.3   When to Use Sampling Methods

Implicit filtering, and the other methods that are derived from coordinate search, are best used in cases where f is either not smooth, not everywhere defined, discontinuous, or stochastic. Such problems arise, for example, when internal iterations fail to converge, when the objective function depends on stochastic models, or when internal switches in the code for the objective function lead to discontinuities. Sampling methods can also be profitably used when derivatives of f are too costly to obtain. The motivating examples for the construction of implicit filtering were problems in which f was a smooth function corrupted by low-amplitude, high-frequency noise (e.g., truncation error in a simulator, stochastic or random computations within the function [123] or table lookup based on noisy data [124, 131, 132]) or which was not defined (i.e., the code for computing f failed) at many points in the nominal design space Ω.

Deterministic sampling methods, whether interpolatory, pure sampling, or a hybrid, require some underlying structure to work well, as the analysis of these methods makes clear [7, 38, 75, 88]. A problem with an objective function which is violently oscillatory is better attacked with a method which requires no structure, such as a genetic algorithm [41, 67, 120] or simulated annealing [1, 85, 128].

One way to decide if a method like implicit filtering is appropriate for your problem is to examine an **optimization landscape**, which is a plot of f against two of the independent variables. The image on the cover of this book is an optimization landscape from the problem from [28, 87], which we discuss in Chapter 10. That landscape has much in common with the one in Figure 1.1, which is taken from [131], the first paper about implicit filtering.

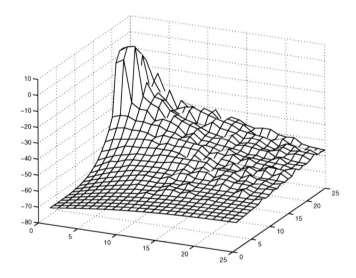

Figure 1.1. *A noisy landscape.*

Both landscapes show multiple local minima and high-frequency, low-amplitude oscillations. These features could easily confound a gradient-based local optimization algorithm. However, in both cases it is clear visually where the optimum should be, or at least where the function value is low. These two features are important indicators that implicit filtering is a good choice.

The two landscapes also illustrate important points by their differences. The landscape in Figure 1.1 has its oscillations and local minima near the peak and away from the optimum. This, in fact, was the motivation for the entire algorithm. The quasi-Newton iteration would do well in the smooth region, which is near the optimum, while the search would step over the local minima and the oscillations when far from optimality.

The landscape of the cover, on the other hand, is oscillatory near the low values of the objective function and, more importantly, has gaps both far from optimality and, as well as one can determine visually, right at the low point on the surface. These gaps reflect variables at which the function does not return a value. Such gaps are a significant problem for fully interpolatory methods, but not as much for search algorithms. The model from [28, 47, 87], which generated the cover figure, and the one from [123, 124, 131, 132], which motivated the development of implicit filtering, are stochastic. This means that there is randomness in the evaluation of f, and hence f is unlikely to return the same value twice for the same input. Algorithms for these problems must be prepared for the possibility that $f(x) \neq f(x)$!

imfil.m can also be used as a multifidelity solver. If you can control the accuracy and cost of evaluating f, you can couple that control to the difference increment in **imfil.m**, reducing the increment as you refine the accuracy of f. This application of **imfil.m** is related to the methods from [24, 25], and we give an example in § 8.3.

The case studies in Part IV illustrate some situations for which sampling methods are useful. One is when the objective function or least squares residual is computed with a simulator and the error tolerances in the simulator are large. Chapter 8 is about an artificial problem [9, 75] which could be treated as a smooth problem were we to reduce the tolerances in the simulator. We include it because the source code is short enough to explain in detail and to allow us to explore several options in **imfil.m**.

The other two case studies are based on recent research papers. The codes for the functions were written by others and should be treated as "black boxes." The problem in Chapter 10 is from water resources policy. There are gaps in the landscape, randomness in the function evaluation, and the function is very noisy [28, 47, 87]. The function evaluation is costly and using parallelism is straightforward.

Chapter 9 is a problem from hydrology [53, 97, 98]. We use an updated version of the implementation from [53] as an example of a problem with discontinuities and to show how to use **imfil.m** for problems where the internal simulator communicates with the external world by file I/O. This makes parallelism challenging.

1.4   When to Avoid Sampling Methods

We will repeat our caveat about smooth problems. It's a poor idea to use sampling methods if you can efficiently compute gradients. While we offer the `smooth_problem` option (see § 6.6.4), we discourage its use unless your function is **scale-aware** (see § 6.6.3) and you can directly control the accuracy and cost of the function evaluation.

Keep in mind that modern gradient-based codes can solve more general problems than the bound constrained problems which **imfil.m** is built to solve. In the classical nonlinear programming problem [51, 56, 108] f is a smooth (i.e., twice Lipschitz continuously differentiable) function and the nominal design space Ω can be described by smooth inequality constraints, i.e.,

$$\Omega_C = \{x \in R^N \,|\, c_i(x) \leq 0, 1 \leq i \leq P\}, \qquad (1.3)$$

where $c_i : R^N \rightarrow R$ is continuously differentiable. There are several good gradient-based methods and codes for solving this classical problem [13, 14, 20, 34, 55, 129]. Sampling methods such as implicit filtering are not among them, and one should use a gradient-based code for such problems.

Chapter 3 is a review of a few traditional methods which are particularly relevant to implicit filtering and its implementation in **imfil.m**.

Chapter 2

Getting Started with imfil.m

In this chapter we give a brief description of what **imfil.m** does, how to install it, and what the computing environment should be. We will illustrate its use for two simple problems. We do not describe all the options, the details of the algorithms, or the ways to extend the code on your own. We will do that in the later chapters.

2.1 Computing Environment and Installation

In order to use **imfil.m**, you will need to get the software, put the codes in your MATLAB path, and be running a recent version of MATLAB. We have tested **imfil.m** on versions 6.5 and higher. Higher is better, especially if you want to use the parallel toolbox.

imfil.m uses very little memory on its own. The codes which define your problem may use much more. MATLAB will complain if it runs out of memory, which is less likely if you run version 7.5 or later.

Installation is easy. Download the MATLAB files for **imfil.m** from

http://www.siam.org/books/se23/

Then put **imfil.m** and **imfil_optset.m** in a directory and put that directory in your MATLAB path.

2.2 What imfil.m does

As we said in Chapter 1, implicit filtering is a sampling method. We start our account of the algorithm with a description of the sampling strategy. Implicit filtering's samples are arranged on a stencil, and it is important to understand how that stencil is built. We begin with a current iterate x_c and the value of the function $f(x_c)$. Then, the default algorithm is to sample the $2N$ points

$$x_c \pm h v_i, \quad 1 \le i \le N,$$

9

where

$$v_i = (L_i - U_i)e_i,$$

e_i is the unit vector in the ith coordinate direction, and h, the **scale**, varies as the optimization progresses. The default sequence of scales is

$$\{2^{-n}\}_{n=\texttt{scalestart}}^{\texttt{scaledepth}}.$$

The algorithmic parameters `scaledepth` and `scalestart` can be changed from the defaults of 7 and 1 with the `imfil_optset` command. The optimization will terminate when the sequence of scales has been exhausted.

   imfil.m uses the values of f on the stencil in several ways, one of which is to construct a difference gradient and use that in a quasi-Newton method (see Chapters 3 and 4 for the details). **imfil.m** reports results after each quasi-Newton iteration is complete. When the supply of scales has been exhausted, the optimization terminates.

   imfil.m scales the bounds by changing variables so that $L_i = 0$ and $U_i = 1$ for all i. Scaling helps **imfil.m** take steps of relatively equal size in all the variables. **You do not have to scale the variables. imfil.m does that for you.** The scaling of x is transparent to you unless you use the `executive_function` (§ 7.5) or `explore_function` options (see Chapter 7).

2.2.1 Constraints

Implicit filtering is able to respond to the function's failure to return a value. When this happens, we say that a **hidden constraint** has been violated. **imfil.m** treats a point in Ω for which f has no value as missing data, and will proceed without the value. Your implementation of f (see § 2.2.3 and § 6.2.2) must communicate a failure to **imfil.m**. We discuss hidden constraints in detail in § 4.4.3.

   Explicit constraints are those that can be evaluated by simply testing the variables and not calling an expensive simulator within f. These are the kinds of constraints one sees in nonlinear programming,

$$c_i(x) \leq 0, \quad 1 \leq i \leq P,$$

where $c_i : R^N \to R$ and the inequality is understood componentwise. If you have explicit constraints, you must communicate infeasibility to **imfil.m** by signaling failure.

   You could also use a penalty function to inform **imfil.m** about explicit constraints [51, 108]. In this approach one replaces f by

$$f_p(x) = f(x) + p(x),$$

where p is an **penalty function** which measures the deviation from feasibility. For example, the **exact l_1 penalty function** for smooth inequality constraints is

$$p(x) = \frac{1}{\mu} \sum_{i=1}^{P} \max(-c_i(x), 0).$$

Here μ is the **penalty parameter**. Selecting μ requires some thought [108].

Constraints can cause problems for stencil-based sampling methods by hiding descent directions from the stencil, and enriching the set of directions is necessary for convergence theory [7, 92] and useful in practice as well. The `add_new_directions` (§ 7.1), `vstencil` (§ 6.9.1), and `random_stencil` (§ 6.9.2) options to **imfil.m** are three ways to do this.

We give examples in § 4.4.4 and § 7.1. We discuss the theory in § 5.6.

2.2.2   The Budget for the Iteration

The most common way to terminate a sampling algorithm is to assign a **budget** of function evaluations to the optimization and to stop the computation when that budget is exceeded. When the function may fail, keeping track of the budget requires more care, and your code for f must help **imfil.m** with that. One thing to consider, for example, is that sometimes a failed point is significantly cheaper to detect than a complete call to f.

So, at a minimum, you must give **imfil.m** an initial iterate, the objective function, the function value at the initial iterate, the budget, and the bounds. **imfil.m** will return the optimal point x and (optionally) a history of the iteration. You can use the history to evaluate the performance of the algorithm or to understand what has happened if the iteration stagnates.

Assigning the budget can be tricky. If the budget is too large, the iteration will waste function evaluations while making very little progress. A small budget, on the other hand, can clearly hide a good solution. We illustrate the effects of poorly sized budgets in § 8 and § 8.4.

2.2.3   The Objective Function

You must write a MATLAB code for f, which will take as its input $x \in R^N$ and return

- a value $fout = f(x)$,

- a flag $ifail$ to signal a failed evaluation ($ifail = 0$ unless the evaluation fails, if the evaluation fails set $ifail = 1$ and $fout = NaN$), and

- $icount$, an estimate of the cost.

The NaN notation ("not a number") comes from the IEEE floating point standard [70, 110]. We use it to indicate missing data in a way that allows MATLAB to propagate it through the computation.

So, the call to f would look like

```
[fout,ifail,icount]=f(x)
```

If your function never fails and thecost of evaluation is independent of x, you can omit the $ifail$ and $icount$ arguments by setting the `simple_function` option to 1. After doing that you may use a function with only one output argument. Use

```
options=imfil_optset('simple_function',1);
```

and then **imfil.m** will accept

```
fout=f(x),
```

and set *ifail* $= 0$ and *icount* $= 1$. If you use the `parallel` option, **imfil.m** will count the evaluations correctly.

2.3   Basic Usage

At a minimum, **imfil.m** requires the objective function f, the bounds in an $N \times 2$ array, with L in the first column and U in the second, and a budget. **imfil.m** will examine a cumulative cost estimate (which uses *icount*) and terminate the optimization when the budget is exceeded. **imfil.m** will not interrupt an iteration in the middle, so you should expect a modest overshoot in the cost of the optimization. **imfil.m** will also terminate when the list of scales has been exhausted. We describe other ways to terminate the iteration in § 6.10 and § 6.11.

A complete call would look like

```
x=imfil(x0,f,budget,bounds);
```

or, if you want the history of the iteration,

```
[x,histout]=imfil(x0,f,budget,bounds);
```

We will use the `histout` array in the examples in this chapter. You can get more information by asking for the `complete_history` structure. We explain the details of the `histout` array and the `complete_history` structure in § 6.3.1 and § 6.3.2.

Remember that if your objective function is a MATLAB .m file, say, `myfun.m`, you'll use the MATLAB function handle notation (the @ symbol) before the name of the function. Then the call would look like

```
x=imfil(x0,@myfun,budget,bounds);
```

`myfun.m` would have to be either in your MATLAB path on in the current directory. Note that we call the function with a MATLAB function handle, rather than using the name of the file in quotes. The reason for this is accommodation of the way way MATLAB handles optional extra arguments to functions.

The `histout` array is an $IT \times (N+5)$-dimensional array, where IT is simply a counter of the number of times the array is updated. The `histout` array is created after the first function evaluation and updated after each approximate gradient computation. For now we will concentrate on the first two columns, which contain the cumulative number of function evaluations *fcount* and the value of f at the end of the iteration.

2.3.1   Termination

There are two iterations which require termination parameters. The **inner iteration** is the quasi-Newton iteration for each value of h. The **outer iteration** is the implicit filtering iteration. In this section we will explain the default termination criteria and list some other ways to terminate these iterations. The details are in § 6.10.

The inner iteration will terminate

- if the value of f at the current point is smaller than the values elsewhere on the finite difference stencil, a condition we will call **stencil failure**, or

- if the internal termination criteria of the quasi-Newton iteration are satisfied.

One can tune both of these criteria, and a user interested in doing that should look at the details in § 4.2.2 before exploring the options in § 6.11.

The outer iteration, by default, terminates when either

- a budget of calls to the function has been exceeded or

- the list of scales has been exhausted.

The budget is an input argument to **imfil.m**, and this mode of termination usually works well. One can do more, and set various options to terminate the iteration when

- the function value has been decreased to a desired target or

- the variation in the function on the stencil is sufficiently small.

See § 6.10 for the details.

2.4   A Very Simple Example

The files for this example are in the `Examples/Simple_Example` directory of the software collection.

In this section we apply **imfil.m** to a simple example to show you how to set up the data and look at the results. We will minimize

$$f(x_1, x_2) = (x_1^2 + x_2^2) * (1 + .1 * \sin(10 * (x_1 + x_2)))$$

subject to the bound constraints

$$-1 \le x_1, x_2 \le 1.$$

It's not hard to see that the optimal point is $x^* = 0$.

To begin we write a MATLAB .m file for f, which we will call `f_easy.m`. The .m file is

```
function [fv,ifail,icount]=f_easy(x)
% F_EASY
% Simple example of using imfil.m
%
fv=x'*x;
```

```
fv=fv*(1 + .1*sin(10 * (x(1) + x(2) ) ));
%
% This function never fails to return a value
%
ifail=0;
%
% and every call to the function has the same cost.
%
icount=1;
```

Note that we include the *ifail* and *icount* in the output arguments to f_easy.m, even though they are not really needed. We could avoid that by using the simple_function option.

We will use **imfil.m** to minimize *f_easy* and then use the histout array to study the details of the iteration. To use **imfil.m** we will need to specify a budget and an initial iterate, which in this example are

$$x_0 = (.5, .5)^T \quad \text{and} \quad budget = 40.$$

Our code driver_easy.m runs **imfil.m** and then prints the first two columns of the histout array.

```
function [x,histout]=driver_easy;
% DRIVER_EASY
% Minimize f_easy with imfil.m
%
% Set the bounds, budget, and initial iterate.
bounds=[-1, 1; -1 1];
budget=40;
x0=[.5,.5]';
%
% Call imfil.
%
[x,histout]=imfil(x0,@f_easy,budget,bounds);
%
% Use the first two columns of the histout array to examine the
% progress of the iteration.
%
histout(:,1:2)
```

The output directly from MATLAB is

```
1.0000e+00    4.7280e-01
3.0000e+00    4.7280e-01
8.0000e+00    4.7280e-01
1.5000e+01    2.6572e-01
2.0000e+01    9.6363e-04
2.5000e+01    9.6363e-04
```

```
3.0000e+01    9.6363e-04
3.5000e+01    9.6363e-04
4.0000e+01    5.7334e-04
4.5000e+01    1.2430e-04
```

The call to **imfil.m** returns

$$x = (8.8 \times 10^{-3}, -6.8 \times 10^{-3})^T$$

as the solution.

The first column is the function evaluation counter, the second the value of the function, and the third the norm of the approximation of the gradient **imfil.m** computes using the function values on the stencil (see § 4.2). One might think that the function evaluation counter should increase by at least four with each iteration, since the stencil has four points. However, if a point in the stencil is infeasible, as two are in the first iteration, the evaluation is skipped. Hence on the first iteration **imfil.m** reports the function value at the initial iterate and the norm of an approximate gradient based on three points (the initial iterate and the two feasible points in the stencil).

We see a decrease in the function in the second column in the early phase of the iteration, as one would expect. Note also that there is a middle part of the iteration where no visible progress is made. This "flat spot" is, unfortunately, common in sampling methods. At the end, the function decreases again. Had we terminated after 20 iterations, we would have missed this improvement. If we increase the budget and the number of scales (see Chapter 8 and § 2.5.4) we'd see further improvement. The `histout` array for this computation indicates the progress.

Note that the iteration terminated over budget. The reason for this is that the function evaluation counter is compared to the budget only after each iteration, so one may expect to exceed the budget by a bit.

2.5   Setting Options

You can set several algorithmic parameters with the `imfil_optset` options command. Many of these are rarely needed or are intended for the specialist. We will discuss only the most useful and important in this section. We will explain the details for all the options in Chapter 6.

If you want to accept the default options, you need do nothing. If you want to explicitly modify the default options structure, you can get if from the `imfil_optset` command by calling that command with no arguments:

```
options=imfil_optset;
```

You need only do this once; additional calls to `imfil_optset` will update the the options structure you've already created. For example, if you want to change `scalestart` to 3 and `scaledepth` to 10, you could call `imfil_optset` three times prior to the call **imfil.m**:

```
options=imfil_optset;
options=imfil_optset('scalestart',3,options);
options=imfil_optset('scaledepth',10,options);
```

You can also put all three of the calls to `imfil_optset` in the code fragment above on a single line

```
options=imfil_optset('scalestart',3,'scaledepth',10);
```

If you want to change an existing set of options, you would add the name of the options structure to the `imfil_optset` command. For example, to change `scaledepth` from 10 to 8, in the options structure you created with the call to `imfil_optset` above, the call would be

```
options=imfil_optset('scaledepth',8,options);
```

You might try to modify `driver_easy.m` by increasing the budget and the number of scales. If you change the call to **imfil.m** to

```
options=imfil_optset('scaledepth,20);
bounds=[-1, 1; -1 1];
budget=100;
x0=[.5,.5]';
%
% Call imfil.
%
[x,histout,complete_history]=imfil(x0,@f_easy,budget,bounds,options);
```

you will see a smaller function value and one more flat spot. You might also try the `smooth_problem` (see § 6.6.4) option.

2.5.1   Nonlinear Least Squares

Many problems, such as the example in Chapter 8, are best formulated as nonlinear least squares problems, where F returns an vector of residuals in R^M and the function to be minimized is

$$f(x) = \|F(x)\|^2/2 = F(x)^T F(x)/2. \tag{2.1}$$

You can tell **imfil.m** that your problem is a nonlinear least squares problem by setting the `least_squares` option to 1 with the command

```
options=imfil_optset('least_squares',1);
```

If you do this, you need to write your function so that $F \in R^M$ is returned. **imfil.m** will construct $f(x) = F(x)^T F(x)/2$ for you. The optimization method is also tuned to a nonlinear least squares computation, and the underlying method is a damped finite difference Gauss–Newton iteration [44, 75]. We refer the reader to § 3.9 for the details of the Gauss–Newton implementation. Chapter 8 has a simple nonlinear least squares example.

2.5.2 Parallel Computing

The **parallel** option tells **imfil.m** that f can be called with multiple arguments and will return a matrix whose columns are the values of f, *ifail*, and *icount*. So if x is an $N \times P$ array of P arguments to f and **parallel** is set to 1, a call to $f(x)$ will return three $1 \times P$ vectors of values and flags. It is your responsibility to write f to do the parallel evaluation in an efficient way. Our example of a parallel call in § 8.2 shows how **imfil.m** responds to

```
options=imfil_optset('parallel',1);
```

If you are solving a nonlinear least squares problem, where a call to f returns an $M \times 1$ column vector, your parallel function should return an $M \times P$ array of residual values as well as vectors *iflag* and *icount*. The parallel algorithm is not the same as the serial method because all the line search possibilities are examined at the same time (see § 6.7 for the details). One implication of this is that more function evaluations can be used even if the final result is the same as in the serial case and the total runtime is significantly less. One should interpret graphs like Figure 8.1 with care when one does the function evaluations in parallel. The default is *parallel* = 0.

The latest versions of MATLAB support some parallelism. The MATLAB parallel computing toolbox **matlabpool** command lets you build a pool of "workers" or "labs," which are separate copies of MATLAB running on each core of a multicore computer. The toolbox also provides the **parfor** loop. A **parfor** loop executes each iteration of the loop on a separate worker, if an idle worker is available. For example, suppose you have an 8 core computer and want to evaluate f at several points $\{x_i\}_{i=1}^{k} \subset R^1$; you might do something like

```
parfor i=1:k
   f(i) = f(x(i));
end
```

after calling **matlabpool(8)** **once** for your MATLAB session. This, in fact, is how one would use **parfor**, but you must pay attention to global variables and memory conflicts among statements in the loop. The examples in Part IV use the **parfor** loop.

Here is an example of a MATLAB session which calls **f_easy** 16 times in parallel on an 8 core computer. After opening MATLAB we begin with a **matlabpool** command to open 8 labs:

```
>> matlabpool(8)
Starting matlabpool using the 'local' configuration
... connected to 8 labs.
```

If you use **matlabpool** while labs are open from a previous call, MATLAB will complain, and you should close all open labs with **matlabpool close**. Our code **parallel_easy** then uses a **parfor** loop to evaluate **f_easy** at some random points and print the results. The codes are in the subdirectory **Examples/Simple_Example** in the software collection.

```
% PARALLEL EASY
%
% A simple matlab parfor example.
% You must call matlabpool if you want this to run in parallel.
%
x=rand(2,16);
f=zeros(16,1);
parfor i=1:16
   f(i) = feval(@f_easy,x(:,i));
end
f
```

The `parallel_easy` script could easily be converted into a parallel version of
`f_easy`:

```
function [fv, ifail, icount]=f_easy_p(x)
% F_EASY_P
%
% A parallel version of f_easy.
% You must call matlabpool if you want this to run in parallel.
%
%function [fv, ifail, icount]=f_easy_p(x)
%
% fv must be a ROW vector. This makes scalar optimization consistent
% with what imfil does for nonlinear least squares.
%
% If you make fv a column vector, you will get some very interesting
% error messages.
%
[nr,nc]=size(x);
fv=zeros(1,nc);
%
% Like f_easy, this function never fails and all calls to f have the
% same cost. However, we have to count the number of calls.
%
ifail=zeros(nc,1);
icount=nc*ones(nc,1);
%
parfor i=1:nc
   fv(i) = feval(@f_easy,x(:,i));
end
```

MATLAB does the sensible thing if you don't have the parallel toolbox and simply
executes a `for` loop. This means that `f_easy_p` will simulate parallel execution even
if it actually works in serial mode.

To change `driver_easy.m` into a code which calls **imfil.m** in parallel, we
change the call to **imfil.m** to

```
%
% Turn the parallel option on.
%
options=imfil_optset('parallel',1);
%
% Call imfil.
%
[x,histout,complete_history]=imfil(x0,@f_easy_p,budget,bounds,options);
%
```

to build `driver_easy_p.m`, the parallel version of `driver_easy.m`.

```
1.0000e+00    4.7280e-01
3.0000e+00    4.7280e-01
8.0000e+00    4.7280e-01
1.6000e+01    7.3599e-03
2.1000e+01    7.3599e-03
2.6000e+01    7.3599e-03
3.1000e+01    7.3599e-03
3.9000e+01    1.5944e-05
4.4000e+01    1.5944e-05
```

The output is a little different from that of the serial code because the parallel and serial versions are slightly different algorithms (see § 6.7):

There are also some very useful resources in the **MATLAB Central File Exchange**. This is a software repository maintained by The MathWorks at

http://www.mathworks.com/matlabcentral/fileexchange/

MULTICORE [19] is a package that lets you use multiple cores with MATLAB. Each core runs its own copy of MATLAB. The package moves data between cores with file I/O, an approach with can slow down the computation if function calls are very inexpensive. The MATLAB `parfor` construct uses memory for the interprocess communication and is significantly faster. However, MULTICORE is free. The software associated with [82] has the `pRUN` program, which allows you to run the same MATLAB code on multiple processors. These approaches do not support fine-grained parallelism (i.e., the use of many processors to speed up the internal computations within f), but should work well for very expensive function evaluations.

2.5.3 Scaling f

If the values of $|f|$ are very small or very large, the quality of the difference gradient which **imfil.m** uses in its search can be poor. **imfil.m** attempts to solve this problem by **scaling** the function by dividing it by the size of a "typical value." Unless you tell **imfil.m** otherwise, this value is 1.2 times the absolute value of the value at the initial iterate.

You can change this by setting the `fscale` option. Setting `fscale` to a negative value will tell **imfil.m** to use $|fscale| \times |f(x_0)|$ as the typical value for f. Setting `fscale` to a positive value will tell **imfil.m** to use *fscale* as the typical value. If you blunder and set *fscale* $= 0$, **imfil.m** will restore the default. If $f(x_0) = 0$, then **imfil.m** will set *fscale* to 1. See § 6.5.1 for more details on `fscale` and its role in **imfil.m**.

Scaling f to order 1 means that we can compare the variation in f (or the change in f from one iteration to the next) to a tolerance (which may depend on the scale) and make a termination decision. See § 6.10 for the details and § 8.4.1 for an example.

2.5.4 Changing the Scales

imfil.m uses a stencil that is built from the bounds. If your current point is x_c, **imfil.m**'s default behavior is to sample the $2N$ points

$$x_c \pm h(L_i - U_i)e_i, \quad 1 \leq i \leq N, \tag{2.2}$$

where e_i is the unit vector in the ith coordinate direction, and h, the **scale**, varies as the optimization progresses. Implicit in the definition of the stencil (2.2) is the finiteness of the bounds.

The sequence of scale is

$$\{2^{-n}\}_{n=scalestart}^{scaledepth}.$$

`scalestart` and `scaledepth` can be changed with the options command. The defaults are *scalestart* $= 1$ and *scaledepth* $= 7$. You can use your own array of scales with the option.

2.5.5 Looking at the Iteration History

You have already seen how the `histout` array can be used to examine the performance of the iteration. **imfil.m** maintains an internal `complete_history` structure which contains the entire history of the iteration (see § 6.3.2). You can access that data either as an optional output argument to **imfil.m** or within the iteration by using one of the advanced options (see Chapter 7).

2.5.6 Scale-Aware Functions

Your function may be able to adjust its own accuracy or resolution. In this case we will say that your function is **scale-aware**. One example of this possibility is if the tolerance in a solver can be reduced as the scale is reduced. This is the way in which **imfil.m** can be used as a multifidelity solver. If your function has this capability, you may enable communication between **imfil.m** and the function call by adding the scale as an extra argument to f, making the call look like

`[fout,ifail,icount]=f(x,h)`

You must tell **imfil.m** that f is taking the extra argument by setting the `scale_aware` option to 1, the default is 0. See § 6.6.3 and § 8.3 for examples.

2.6 Passing Data to the Function

You may need to pass data from your calling program directly to f. For example, the data for a nonlinear least squares problem is part of the least squares residual, but you may not want to hard code that data into the function. **imfil.m** permits an optional final argument which you may use for that purpose. The calling sequence looks like

```
[x,histout]=imfil(x0,f,budget,bounds,options,extra_data);
```

Here `extra_data` can be an array, a function handle, or a structure (see § 6.8 for more details). Chapters 8, 9, and 10 provide examples of how this can be used.

You might think that you could use MATLAB global variables for this purpose. However, global variables can cause problems with parallel computing in MATLAB, and we recommend that you avoid them.

Chapter 3

Notation and Preliminaries

Complete understanding of implicit filtering requires some knowledge of numerical linear algebra, calculus, and classical optimization theory. In this chapter we review this material and express it in a way we can use in the algorithmic and theory chapters which follow this one.

3.1 Numerical Linear Algebra

We begin with a brief review of some important ideas from numerical linear algebra. Books like [42, 60, 101, 122, 127] cover this material in depth. The important topics are Gaussian elimination for linear equations and the solution of overdetermined linear least squares problems. Our review is intended to highlight the ideas which are important in **imfil.m**, not to provide a detailed review.

In this book vectors are column vectors, R^N is the space of real vectors of length N, and $(x)_i$ is the ith component of a vector x. We use this notation, as we did in [75], to distinguish the i component from the i element of a sequence of iterations. We will consistently use the Euclidean norm

$$\|x\| = \sqrt{\sum_{i=1}^{N}(x)_i^2} \tag{3.1}$$

and its induced matrix norm

$$\|A\| = \max_{x \neq 0} \frac{\|Ax\|}{\|x\|}, \tag{3.2}$$

where A is an $M \times N$ matrix.

The **condition number** of a square matrix A is

$$\kappa(A) = \|A\|\|A^{-1}\|. \tag{3.3}$$

We say the matrix A is **well-conditioned** if $\kappa(A)$ is not too large. What we mean by large will depend on the context. The condition number depends on one's choice of matrix norm, which in our case is the Euclidean norm. If A is very ill-conditioned, the numerical solution of $Ax = b$ will be unreliable.

3.1.1   The LU Factorization and Gaussian Elimination

The Gaussian elimination algorithm one sees in elementary courses is realized in modern software as a matrix decomposition. The MATLAB `lu` command performs Gaussian elimination and encodes the result as a pair of (column-permuted) triangular matrices which factor A as $A = LU$, where L is a column-permuted lower triangular matrix and U is an upper triangular matrix. The MATLAB backslash operator solves $Ax = b$ via the command

```
b = A\x
```

and uses Gaussian elimination to do that. Internally the factorization is done first and then followed by two triangular solves.

3.1.2   The Cholesky Factorization

Positive definite matrices play a role in both theory and algorithm design for optimization. If a matrix is symmetric and positive definite, we can save both time and storage in its LU factorization. The Cholesky factorization is $A = LL^T$, where L is lower triangular with positive diagonal elements.

Definition 3.1. *An $N \times N$ matrix A is* **positive semidefinite** *if $x^T Ax \geq 0$ for all $x \in R^N$. A is* **positive definite** *if $x^T Ax > 0$ for all $x \in R^N, x \neq 0$. If A has both positive and negative eigenvalues, we say A is* **indefinite**. *If A is symmetric and positive definite, we will say A is* **spd***.*

The linear equations one has to solve within **imfil.m** often have positive definite coefficient matrices. The MATLAB backslash operator will use a Cholesky factorization if the coefficient matrix is spd.

3.1.3   Linear Least Squares

imfil.m will solve nonlinear least squares problems. The standard approach [44, 75] is to use a Newton-like approach and model the nonlinear problem by a linear least squares problem, solve the model problem, and use the solution of the model problem as the next iteration. In this section we discuss standard methods [42, 60] for solving linear least squares problems.

The **overdetermined** linear least squares problem [15, 42, 60] is to find $x \in R^N$, which minimizes $\|Ax - b\|$, where A is an $M \times N$ matrix, $b \in R^M$, and $M \geq N$. We will express this as

$$\min_{x} \|Ax - b\|. \tag{3.4}$$

We will assume throughout this book that A has **full column rank**, i.e., the columns of A are linearly independent.

One simple way to solve (3.4) is to form the **normal equations**

$$A^T A x = A^T b, \tag{3.5}$$

and solve that system with a Cholesky factorization. Since A has full column rank, $A^T A$ is spd and this approach will in principle work fine, and $x^* = (A^T A)^{-1} A^T b$ is the unique solution.

The matrix $A^T A$ can be very ill-conditioned, and it's generally better to solve (3.4) with a **QR factorization**. Recall that the QR factorization of A decomposes

$$A = QR,$$

where Q is an $M \times N$ matrix with orthonormal columns and R is an $N \times N$ upper triangular matrix. R is nonsingular because A has full column rank. MATLAB computes this with the "economy size" decomposition $[Q, R] = qr(A,0)$. With the QR factorization in hand we note that

$$\|Ax - b\| = \|Q^T (Ax - b)\| = \|Rx - Q^T b\|$$

is minimized when

$$x^* = R^{-1} Q^T b. \tag{3.6}$$

Hence there is no need to form or solve the normal equations. Also

$$\|Ax^* - b\| = \|(I - QQ^T)b\|.$$

The QR method and the normal equations approach are connected in a simple way, since

$$A^T A = (QR)^T (QR) = R^T (Q^T Q) R = R^T R, \tag{3.7}$$

we see that $A^T A = R^T R$ could be used instead of the Cholesky decomposition to solve the normal equations. While this connection is not useful as a solver, it will be helpful when we consider bound constrained nonlinear least squares problems in § 3.9. One can also understand the advantage of using the QR factorization over the normal equations approach by taking condition numbers in (3.7) to see that $\kappa(A^T A) = \kappa(R)^2 = \kappa(A)^2$.

We can quantify the conditioning of a linear least squares problem with the **singular value decomposition** (SVD). To begin with we assume that $M \geq N$ and will use the "thin" or "economy size" SVD throughout this book. The MATLAB command $svd(A,0)$ computes this form of the SVD. Here we factor A as

$$A = \mathcal{U} \Sigma \mathcal{V}^T, \tag{3.8}$$

where $\mathcal{U}$ is an $M \times N$ matrix with orthogonal columns, Σ is a diagonal matrix with the **singular values** (square roots of the eigenvalues of $A^T A$) on the diagonal, and $\mathcal{V}$ is an $N \times N$ orthogonal matrix ($\mathcal{V}^T \mathcal{V} = \mathcal{V}\mathcal{V}^T = I$). The columns of $\mathcal{U}$ and $\mathcal{V}$ are called the left and right **singular vectors**. The solution of (3.4) is

$$x = \mathcal{V} \Sigma^{-1} \mathcal{U}^T b.$$

If $N > M$, then, again following [42], we form the SVD of A^T as above and then take the transpose of that SVD. In this way Σ is always a square diagonal dimension of size $N \times N$ if $M \geq N$ and size $M \times M$ if $N > M$.

The singular values $\{\sigma_i\}_{i=1}^{N}$ are decreasing. The condition number of the linear least squares problem is the ratio σ_1/σ_N of largest to smallest singular values. When the condition number is large the solution is highly sensitive to perturbations in A and b and the results are unreliable. Such problems are called **ill-posed** and must be **regularized** to be solved reliably.

One approach to regularization is to discard small singular values. If, for example, $\{\sigma_i\}_{i=\ell+1}^{N}$ are significantly smaller that $\{\sigma_i\}_{i=1}^{\ell}$, one can ignore the directions $\{u_i\}_{i=\ell+1}^{N}$ in the formulation of (3.4) by setting $\{\sigma_i\}_{i=\ell+1}^{N}$ to zero to obtain

$$\Sigma_S = \mathrm{diag}(\sigma_1, \ldots, \sigma_\ell, 0 \ldots 0)$$

and approximating the solution of (3.4) by

$$x = \mathcal{V}\Sigma_S^\dagger \mathcal{U}^T b.$$

Here the **pseudoinverse** $\Sigma_S^\dagger$ of Σ_S is

$$\Sigma_S^\dagger = \mathrm{diag}(\sigma_1^{-1}, \ldots, \sigma_\ell^{-1}, 0 \ldots 0) \tag{3.9}$$

and that of A is

$$A^\dagger = \mathcal{V}\Sigma_S^\dagger \mathcal{U}^T. \tag{3.10}$$

This approach to regularization is related to the method of principal component analysis or empirical orthogonal functions [73, 111] in statistics.

If you have small singular values in a data-fitting problem, as many applications of nonlinear least squares are, using a pseudoinverse to handle small singular values is very sensitive to errors [71, 72, 121]. The method of subset selection [60, 62] prunes the variables in a clever way, allowing you to identify those variables which are hardest to fit. You can (and should) set them to nominal values. The least squares problem for the remaining variables will be much better conditioned.

We will also need to compute the SVD with $N > M$ as part of implicit filtering's approximation of derivatives. Following [42] we define the SVD of an $M \times N$ matrix A with $N > M$ as the transpose of the "economy" SVD of A^T.

3.2   Preliminaries from Calculus

We will denote maps from $R^N \to R^M$ by F. In the special case of real-valued maps (i.e., $M = 1$) we will use lowercase f. This distinction is particularly important when we discuss nonlinear least squares problems.

The **Jacobian** or **Jacobian matrix** of a map $F : R^N \to R^M$ is the $M \times N$ matrix F' whose entries are

$$F'_{ij}(x) = \partial F_i / \partial(x)_j.$$

Let f be a real-valued function of a vector variable $x \in R^N$. For $x \in R^N$ we let $\nabla f(x) \in R^N$ denote the **gradient** of f,

$$\nabla f(x) = (\partial f / \partial(x)_1, \ldots, \partial f / \partial(x)_N)^T,$$

when it exists. Note that ∇f is the transpose of the Jacobian of f.

We let $\nabla^2 f$ denote the **Hessian** of f,

$$(\nabla^2 f)_{ij} = \partial^2 f / \partial(x)_i \partial(x)_j,$$

when it exists. Note that $\nabla^2 f$ is the Jacobian of ∇f. However, $\nabla^2 f$ has more structure than a Jacobian for a general nonlinear function. If f is twice continuously differentiable, then the Hessian is symmetric $((\nabla^2 f)_{ij} = (\nabla^2 f)_{ji})$ by equality of mixed partial derivatives [116].

We say a function $F : R^N \to R^M$ is **Lipschitz continuous** with **Lipschitz constant** L if

$$\|F(x) - F(y)\| \le L\|x - y\| \tag{3.11}$$

for all x, y, where the two norms depend on the domain and range of F. We will often assume that F is **Lipschitz continuously differentiable**, which means that the Jacobian F' is a Lipschitz continuous function of its argument. A real-valued function f is **twice Lipschitz continuously differentiable** if $\nabla^2 f$ is a Lipschitz continuous matrix-valued function of x.

3.2.1 The Fundamental Theorem of Calculus and Taylor's Theorem

We will need to express the **fundamental theorem of calculus**, in several ways.

Theorem 3.2. *Let $F : R^M \to R^N$ be continuously differentiable on the line segment between points x^* and $x = x^* + e$ in R^N. Then*

$$F(x) = F(x^*) + \int_0^1 F'(x^* + te)e \, dt. \tag{3.12}$$

If f is a twice continuously differentiable real-valued function on this line segment, then

$$f(x) = f(x^*) + \int_0^1 \nabla f(x^* + te)^T e \, dt$$

and

$$\nabla f(x) = \nabla f(x^*) + \int_0^1 \nabla^2 f(x^* + te)e \, dt.$$

We will use **Taylor's theorem** throughout the book. We will state the theorem twice, once for vector-valued F and again for real-valued f.

Theorem 3.3. *Let $F : R^M \to R^N$ be Lipschitz continuously differentiable in a neighborhood of a point $x^* \in R^N$. Then for $e \in R^N$ and $\|e\|$ sufficiently small*

$$F(x^* + e) = F(x^*) + F'(x^*)e + O(\|e\|^2). \tag{3.13}$$

Theorem 3.4. *Let f be twice Lipschitz continuously differentiable in a neighborhood of a point $x^* \in R^N$. Then for $e \in R^N$ and $\|e\|$ sufficiently small*

$$f(x^* + e) = f(x^*) + \nabla f(x^*)^T e + e^T \nabla^2 f(x^*)e/2 + O(\|e\|^3). \qquad (3.14)$$

Taylor's theorem constructs a **local model** or **local surrogate** of a vector-valued function F near x^* using the values of F and F' and of a real-valued function near x^* using the values of f, ∇f, and $\nabla^2 f$. Methods like Newton's method, steepest descent, and implicit filtering can be viewed as method which build a local model about the current iteration, minimize that local model, and use the minimizer of the local model as the next iterate.

3.3 Unconstrained and Bound Constrained Optimization

Implicit filtering is built upon classical optimization methods such as steepest descent, gradient projection, and quasi-Newton acceleration. These topics are covered in many modern optimization books [75, 108]. This section reviews these topics from the point of view of [75].

The **unconstrained optimization** problem is to minimize a real-valued **objective function** f of N variables. By this we mean to find a **local minimizer**, that is, a point x^* such that

$$f(x^*) \leq f(x) \text{ for all } x \text{ near } x^*. \qquad (3.15)$$

It is standard to express this problem as

$$\min_x f(x) \qquad (3.16)$$

or to say that we seek to solve the problem min f. The understanding is that (3.15) means that we seek a local minimizer. We call $f(x^*)$ the **minimum** or **minimum value**. The problem (3.16) is called **unconstrained** because we put no conditions on x and assume that the objective function is defined for all $x \in R^N$.

The **bound constrained optimization** problem is to minimize an objective function f over a hyperrectangle in R^N. A hyperrectangle is a set of the form

$$\Omega = \{x \in R^N \mid L_i \leq (x)_i \leq U_i\}. \qquad (3.17)$$

Recall that $(x)_i$ is the ith component of the vector x. We express the bound constrained problem as

$$\min_{x \in \Omega} f(x), \qquad (3.18)$$

and the notation means that we seek a local minimizer, i.e., $x^* \in \Omega$ such that

$$f(x^*) \leq f(x) \text{ for all } x \in \Omega \text{ near } x^*.$$

3.4   Necessary and Sufficient Conditions for Optimality

The necessary conditions for optimality relate optimality to the solution of a non-linear equation, for which methods like Newton's method [44, 74, 77, 109] are available. One must take care to avoid converging to a local maximum, of course, but the necessary conditions are the first step toward the use of local surrogates in minimization.

3.4.1   Unconstrained Problems

The necessary conditions for unconstrained optimization are simply that $\nabla f(x^*) = 0$ (**stationarity**) and that $\nabla^2 f(x^*)$ be positive semidefinite (i.e., no direction clearly points downhill). A nonlinear equation code may well find a solution of $\nabla f(x) = 0$, but if there is no attempt to examine $\nabla^2 f$, the iteration may also converge to a local maximizer. The necessary conditions are not sufficient, for example, the function $f(x) = x^3$ satisfies the necessary conditions at $x = 0$, but 0 is not a local minimizer.

Theorem 3.5. *Let f be twice continuously differentiable and let x^* be a local minimizer of f. Then*

$$\nabla f(x^*) = 0 \tag{3.19}$$

and $\nabla^2 f(x^*)$ is positive semidefinite.

Many iterative methods terminate when (3.19), the **first-order necessary conditions**, are nearly satisfied.

The assumptions of Theorem 3.5 are called the **second-order necessary conditions**. The **second-order sufficiency conditions** are the assumptions of Theorem 3.6.

Theorem 3.6. *Let f be twice continuously differentiable in a neighborhood of x^*. Assume that $\nabla f(x^*) = 0$ and that $\nabla^2 f(x^*)$ is positive definite. Then x^* is a local minimizer of f.*

3.4.2   Bound Constrained Problems

If the solution of a bound constrained problem is in the interior of the feasible set, i.e.,

$$L_i < (x^*)_i < U_i$$

for all i, then there is no difference from the unconstrained case in the necessary or sufficient conditions for optimality.

We will say that the ith constraint is **active** at $x \in \Omega$ if either $(x)_i = L_i$ or $(x)_i = U_i$. If the ith constraint is not active, we will say that it is **inactive**. The set of indices i such that the ith constraint is active (inactive) will be called the set of **active (inactive) indices** at x. The subset $\mathcal{B}$ of **binding constraints** are those active constraints for which

$$(\partial f(x)/\partial(x)_i)((x)_i - (y)_i) < 0$$

for all $y \in \Omega$. Indices not in $\mathcal{B}$ are **nonbinding**. We will write $\mathcal{A}(x)$, $\mathcal{B}(x)$, $\mathcal{I}(x)$, and $\mathcal{N}(x)$ for the active, binding, inactive, and nonbinding sets at x.

The necessary conditions say that x^* is optimal if there is no clear way to reduce f and remain in Ω. For example, if $f(x) = x$ and $x_c = 1$, then $d = -f'(x_c) = -1 \neq 0$, and f can be reduced by moving to the left. If, however, $\Omega = [1, 2]$, then it is not possible to move to the left and remain in Ω, and then $x_c = 1$ will satisfy the first-order necessary conditions we define below.

Definition 3.7. *A point $x^* \in \Omega$ is* **stationary** *for problem* (3.18) *if*

$$\nabla f(x^*)^T(x - x^*) \geq 0 \text{ for all } x \in \Omega. \tag{3.20}$$

As in the unconstrained case, stationary points are said to satisfy the **first-order necessary conditions**.

We can express the first-order necessary conditions in terms of the nonbinding set at x^* [75]. If $i \in \mathcal{N}(x^*)$, then

$$(\nabla f(x^*))_i = 0.$$

If, however, $i \in \mathcal{B}(x^*)$, then $(x^*)_i$ is either L_i or U_i. Moreover, (3.20) holds if and only if $x^* - \lambda \nabla f(x^*)$ is not in the interior of Ω for any $\lambda > 0$, for if it were, we could reduce f and remain in Ω. So, stationarity is equivalent to the nonsmooth nonlinear equation,

$$x^* = \mathcal{P}(x^* - \lambda \nabla f(x^*)) \text{ for all } \lambda \geq 0. \tag{3.21}$$

In (3.21) $\mathcal{P}$ denotes the **projection** onto Ω, that is, the map that takes x into the nearest point (in the l^2 norm) in Ω to x,

$$\mathcal{P}(x) = \max(L, \min(x, U)), \tag{3.22}$$

where the max and min are taken componentwise.

The second-order necessary and sufficient conditions are also more complex than in the unconstrained case. Rather than examine the Hessian, we must pay attention to the binding set as well. The **modified Hessian** helps us do this.

Definition 3.8. *Let f be twice differentiable at $x \in \Omega$. The* **modified Hessian** $\nabla_R^2 f(x)$ *is the matrix whose entries are*

$$(\nabla_R^2 f(x))_{ij} = \begin{cases} \delta_{ij} & \text{if } i \in \mathcal{B}(x) \text{ or } j \in \mathcal{B}(x), \\ (\nabla^2 f(x))_{ij} & \text{otherwise.} \end{cases} \tag{3.23}$$

Here δ_{ij} are the Kronecker delta

$$\delta_{ij} = \begin{cases} 1 & \text{if } i = j, \\ 0 & \text{otherwise.} \end{cases}$$

The modified Hessian is an $N \times N$ matrix, which we use for convenience. The **reduced Hessian** from constrained optimization [108] is composed of the rows and columns of the modified Hessian corresponding to the nonbinding indices. This is a smaller matrix whose size depends on the binding set.

So, now we can state the theorem [11, 75] on necessary conditions for bound constrained problems. The **second-order necessary conditions** are the conclusions of Theorem 3.9.

Theorem 3.9. *Let f be twice continuously differentiable on Ω and let x^* be a local minimizer of f. Then (3.21) holds (i.e., x^* is a stationary point) and $\nabla^2_R f(x^*)$ is positive semidefinite.*

We will present an extended form [93] of the sufficient conditions from [75]. The **second-order sufficient conditions** are the assumptions of Theorem 3.10.

Theorem 3.10. *Let $x^* \in \Omega$ stationary point for problem (3.18). Let f be twice differentiable in a neighborhood of x^* and assume that $\nabla^2_R f(x^*)$ is positive definite. Then x^* is a solution of problem (3.18).*

3.5 Steepest Descent and Gradient Projection Algorithms

Recall that the method of steepest descent for unconstrained minimization seeks local minima of a smooth function f by searching along a ray in the direction of the negative gradient (the steepest descent direction). We will refer to methods that use the gradient to determine a search direction as **gradient-based**.

In this section we will look at the two simplest such methods, the method of steepest descent for unconstrained problems and the gradient projection algorithm [11] for bound constrained problems. Both methods have problems that can make them perform poorly, and the quasi-Newton methods (see § 3.8) are satisfactory remedies for our purposes.

3.5.1 The Method of Steepest Descent

In the steepest descent method for unconstrained problems, we begin with a **current point** x_c, and search for a new point x_+ along the ray from x_c in the direction $-\nabla f(x_c)$. So the new point will have the form

$$x_+ = x_c - \lambda \nabla f(x_c). \tag{3.24}$$

λ is called the **step length** or **step size**. We will use the classical **Armijo rule** [4] to find λ. The process is to begin with $\lambda = 1$ and reduce λ by a constant factor β (a factor of $\beta = 1/2$ is common) until we obtain a **sufficient decrease**, i.e.,

$$f(x_c - \lambda \nabla f(x_c)) - f(x_c) < -\alpha \lambda \|\nabla f(x_c)\|^2. \tag{3.25}$$

If (3.25) holds, then we **accept the step**. In (3.25) α is an algorithmic parameter, typically set to 10^{-4}.

Algorithm steep is a simple algorithm. The iteration terminates successfully if the necessary conditions are nearly satisfied. We express this by asking for small gradients,

$$\|\nabla f(x)\| \leq \tau \tag{3.26}$$

for some small τ. The iteration terminates with failure if a budget of evaluations of ∇f is exceeded. This approach to termination, testing for an approximation of the first-order necessary conditions and holding the iteration to a budget, is a theme of all the algorithms we discuss in this book.

Algorithm 3.1.
steep$(x, f, \tau, kmax)$

    Compute f and ∇f.
    $k = 1$.
    while $k <= kmax$ and $\|\nabla f(x)\| \geq \tau$ **do**
        Find the least integer $m \geq 0$ such that (3.25) holds for $\lambda = \beta^m$.
        $x \leftarrow x - \lambda \nabla f$.
        $k \leftarrow k + 1$.
        Compute f and ∇f.
    end while
    if $\|\nabla f(x)\| < \tau$ **then**
        signal success
    else
        signal failure
    end if

The Armijo rule is an example of a **line search** in which one searches on a ray from x_c in a **descent direction** for f from x_c. A descent direction d is one from which f decreases from x_c, i.e.,

$$\left. \frac{df(x_c + td)}{dt} \right|_{t=0} = \nabla f(x_c)^T d < 0. \tag{3.27}$$

Clearly the steepest descent direction $d = -\nabla f(x_c)$ is a descent direction unless the necessary condition $\nabla f(x_c) = 0$ holds. If d is a descent direction and ∇f is Lipschitz continuous, then one can prove [44, 74, 75, 109] that the line search will eventually succeed, i.e., one will eventually find m such that $\lambda = \beta^m$ satisfies (3.25). If, however, the descent direction is an approximation, for example, a difference approximation of the negative gradient, then the line search may fail.

We will consider a more general form of steepest descent, where the descent directions are based on **quadratic models** of f of the form

$$m(x) = f(x_c) + \nabla f(x_c)^T (x - x_c) + \frac{1}{2}(x - x_c)^T H_c (x - x_c),$$

where the **model Hessian** H_c is spd. We let $d = x - x_c$ be such that $m(x)$ is

minimized. Hence
$$\nabla m(x) = \nabla f(x_c) + H_c(x - x_c) = 0$$
and hence
$$d = -H_c^{-1}\nabla f(x_c). \tag{3.28}$$

We seek to minimize the local model over all of R^N; hence we must insist on a positive definite model Hessian. If H_c has a negative eigenvalue, the eigenvector in that direction is called a **direction of negative curvature** and the quadratic model goes to $-\infty$ in that direction.

The steepest descent direction satisfies (3.28) with $H_c = I$. However, the Newton direction $d = -\nabla^2 f(x)^{-1}\nabla f(x)$ may fail to be a descent direction if x is far from a minimizer because $\nabla^2 f$ may not be spd. Hence, unlike the case for nonlinear equations [74], Newton's method is not a generally good global method, even with a line search, and must be modified (see [48, 54, 56, 117]) to make sure that the model Hessians are spd. We modify Algorithm steep to incorporate computation of the model Hessian and the solution of the linear equation for the search direction and obtain Algorithm steeph below. We must also change the sufficient decrease condition from (3.25) to
$$f(x_c + \lambda d) - f(x_c) < \alpha\lambda\nabla f(x_c)^T d. \tag{3.29}$$

Here, as in (3.25), $\alpha \in (0, 1)$ is an algorithmic parameter. Typically $\alpha = 10^{-4}$.

Algorithm 3.2.
steeph$(x, f, \tau, kmax)$
   Compute f and ∇f
   $k = 1$
   while $k \leq kmax$ and $\|\nabla f(x)\| \geq \tau$ **do**
      Compute the model Hessian H.
      Solve $Hd = -\nabla f(x)$ for the search direction d.
      Find the least integer $m \geq 0$ such that (3.29) holds for $\lambda = \beta^m$.
      $x \leftarrow x + \lambda d$.
      $k \leftarrow k + 1$.
      Compute f and ∇f
   end while
   if $\|\nabla f(x)\| < \tau$ **then**
      signal success
   else
      signal failure
   end if

So, how does it work? Theorem 3.11 is a typical convergence result for optimization and says that if the problem is reasonable (i.e., f is bounded from below and sufficiently smooth) and the model Hessians remain bounded and well-conditioned, then the iteration will drive the gradient to zero, i.e., approach the first-order necessary conditions. This is about all one can expect from the theory, but in practice you can expect to find a local minimum.

Theorem 3.11.   *Let ∇f be Lipschitz continuous. Assume that the matrices H_k are spd and the sequences $\{\|H_k\|\}$ and $\{\|H_k^{-1}\|\}$ are bounded. Then either $f(x_k)$ is unbounded from below or*

$$\lim_{k \to \infty} \nabla f(x_k) = 0, \tag{3.30}$$

and hence any limit point of the sequence of iterates produced by Algorithm steeph *is a stationary point.*

In particular, if $f(x_k)$ is bounded from below and $x_{k_l} \to x^*$ is any convergent subsequence of $\{x_k\}$, then $\nabla f(x^*) = 0$.

Our line search is primitive. In [44, 75] one can find more sophisticated methods which model f on the ray from x_c in the descent direction. Such methods are sequential, and a parallel line search, as we propose in this book, is most easily implemented with the simple approach.

3.5.2   The Gradient Projection Method

The most elementary form of the gradient projection method iteration [11] is the natural analogue of steepest descent for bound constrained problems. The iteration is

$$x_+ = \mathcal{P}(x_c - \lambda \nabla f(x_c)), \tag{3.31}$$

which simply takes a steepest descent step and, in order to keep x_+ feasible, projects it onto

$$\Omega = \{x \in R^N \,|\, L_i \le (x)_i \le U_i\}.$$

If the iteration converges to x^*, we would hope that the projected gradient at x^*

$$x^* - \mathcal{P}(x^* - \nabla f(x^*)) \tag{3.32}$$

would vanish. Under reasonable assumptions, it does. Similar to the steepest descent case, the performance is much better with a good model Hessian. The construction of a useful model Hessian is a bit subtle in this case, so we will begin with the simplest version.

The next step is the Armijo line search. For $\lambda > 0$ define

$$x(\lambda) = \mathcal{P}(x_c - \lambda \nabla f(x_c)).$$

We will adjust the step length λ as we did in the unconstrained case, reducing it until we find sufficient decrease, and then set

$$x_+ = x(\lambda).$$

There are several ways to define sufficient decrease in the bound constrained case. In [75] we use

$$f(x(\lambda)) - f(x_c) \le \frac{-\alpha}{\lambda} \|x_c - x(\lambda)\|^2. \tag{3.33}$$

An alternative from [11, 12] is

$$f(x(\lambda)) - f(x_c) \le -\alpha \nabla f(x_c)^T (x_c - x(\lambda)), \tag{3.34}$$

where

$$x(\lambda) = \mathcal{P}(x_c - \lambda \nabla f(x_c)). \tag{3.35}$$

In (3.34), the parameter α plays the same role as it has throughout this chapter and is again set to 10^{-4}.

Following the approach with the steepest descent method, we will terminate the gradient projection iteration when the projected gradient is small, i.e., when

$$\|x_c - x(1)\| = \|x_c - \mathcal{P}(x_c - \nabla f(x_c))\| < \tau \tag{3.36}$$

for some sufficiently small τ.

The elementary gradient projection algorithm is shown below.

Algorithm 3.3.
gradproj$(x, f, \tau, kmax)$
   $k = 1$.
   Compute f and ∇f; terminate if (3.36) holds.
   while $k \le kmax$ and $\|x - x(1)\| > \tau$ **do**
      Find the least integer m such that (3.34) holds for $\lambda = \beta^m$.
      $x = x(\lambda)$.
      $k = k + 1$.
      Compute f and ∇f.
   end while
   if $\|x - x(1)\| < \tau$ **then**
      signal success
   else
      signal failure
   end if

The convergence of the method depends on identifying the binding constraints. One convergence result [11] is the following theorem.

Theorem 3.12. *Assume that ∇f is Lipschitz continuous. Let $\{x_n\}$ be the sequence generated by the gradient projection method. Then every limit point of the sequence is a stationary point. Moreover, if $\{x_n\}$ converges to a local minimizer x^* which satisfies the sufficient conditions, then there is n_0 such that $\mathcal{B}(x_n) = \mathcal{B}(x^*)$ for all $n \ge n_0$.*

The last sentence in the statement of Theorem 3.12 is the significant result that the binding set is identified in finitely many iterations. Thereafter, the iteration is governed by the theory for unconstrained problems, because a component of x either is frozen at one of the bounds or is free to vary for the remainder of the iteration.

If we wish to use a model Hessian, we must pay attention to the binding constraints and model the modified Hessian instead. Since we cannot know if we are far enough along in the iteration to have identified the active set, we must be

conservative and make sure that we do not overestimate the nonbinding set. The
technical reasons for this approach are explained in [12, 75].

The method from [12] is to underestimate the nonbinding set in a careful way
and therefore maintain a useful and spd approximation to $\nabla^2_R f(x)$. There are many
ways to do this [12, 93]. Here, for $x \in \Omega$ and $0 \le \epsilon < \min(U_i - L_i)/2$, we define
$\mathcal{B}^\epsilon(x)$, the **ϵ-binding set** at x, by

$$\mathcal{B}^\epsilon(x) = \{i \mid U_i - (x)_i \le \epsilon \text{ and } \partial f(x)/\partial(x)_i \le -\sqrt{\epsilon}\}$$
$$\cup \{i \mid (x)_i - L_i \le \epsilon \text{ and } \partial f(x)/\partial(x)_i \ge \sqrt{\epsilon}\} \tag{3.37}$$

and let $\mathcal{N}^\epsilon(x)$, the **ϵ-nonbinding set**, be the complement of $\mathcal{B}^\epsilon(x)$.

We will state and prove a lemma on identification of the binding constraints.
This lemma is needed in the proof of Theorem 3.14 [75, 79, 93] and we will use it
in an important way in § 5.5.

Lemma 3.13. *Let f be Lipschitz continuously differentiable. Let x^* be a local
minimizer which satisfies the second-order sufficiency conditions. For any $x \in \Omega$,
$\epsilon, \alpha > 0$, let*

$$\mathcal{B}^\epsilon(x, \alpha) = \{i \mid U_i - (x)_i \le \epsilon \text{ and } \partial f(x)/\partial(x)_i \le -\alpha\sqrt{\epsilon}\}$$
$$\cup \{i \mid (x)_i - L_i \le \epsilon \text{ and } \partial f(x)/\partial(x)_i \ge \alpha\sqrt{\epsilon}\}. \tag{3.38}$$

Then if ϵ and $\|x - x^*\|$ are sufficiently small,

$$\mathcal{B}^\epsilon(x, \alpha) = \mathcal{B}(x^*).$$

Proof. We let $\|x - x^*\| = \delta$ and will adjust δ as the proof evolves. Let L be
the Lipschitz constant of ∇f.

Suppose $i \in \mathcal{B}(x^*)$ and, without loss of generality, $(x^*)_i = U_i$. Then $\partial f(x^*)/\partial(x)_i$
$= -d_i < 0$. We will assume that $\alpha\sqrt{\epsilon} < d_i$. If

$$\delta < (d_i - \alpha\sqrt{\epsilon})/L,$$

then Lipschitz continuity of ∇f implies that

$$\partial f(x)/\partial(x)_i \le -d_i + L\delta < -\alpha\sqrt{\epsilon}.$$

Hence $i \in \mathcal{B}^\epsilon(x, \alpha)$ and therefore

$$\mathcal{B}(x^*) \subset \mathcal{B}^\epsilon(x, \alpha).$$

Conversely if $i \notin \mathcal{B}(x^*)$, then either

$$\min(U_i - (x^*)_i, (x^*)_i - L_i) = m_i > 0,$$

which implies that $i \notin \mathcal{B}^\epsilon(x, \alpha)$ for any $\epsilon < m_i$, or i is active but not binding. In
the latter case we may assume without loss of generality that

$$(x^*)_i = U_i \quad \text{and} \quad \partial f(x^*)/\partial(x)_i = 0.$$

So if δ is sufficiently small, we may use Lipschitz continuity of ∇f to obtain

$$\partial f(x)/\partial(x)_i > -L\delta.$$

Hence if $\delta < \epsilon$ is sufficiently small, we cannot have

$$\partial f(x)/\partial(x)_i < -\alpha\sqrt{\epsilon},$$

and therefore $i \notin \mathcal{B}^\epsilon(x, \alpha)$. This completes the proof. $\square$

In the discussion in this section we use $\alpha = 1$.

If $\mathcal{S}$ is any set of indices, we define the diagonal matrix $\mathcal{P}_\mathcal{S}$ by

$$(\mathcal{P}_\mathcal{S} x)_{ii} = \begin{cases} 1, & i \in \mathcal{S}, \\ 0, & i \notin \mathcal{S}. \end{cases}$$

Given $0 \leq \epsilon_c < \min(U_i - L_i)/2$, x_c, and an spd matrix H_c, we model $\nabla_R^2 f(x_c)$ with $\mathcal{R}_c$, the matrix with entries

$$\mathcal{R}_c = \mathcal{P}_{\mathcal{B}^{\epsilon_c}} + (I - \mathcal{P}_{\mathcal{B}^{\epsilon_c}})H_c(I - \mathcal{P}_{\mathcal{B}^{\epsilon_c}})$$

$$= \begin{cases} \delta_{ij} & \text{if } i \in \mathcal{B}^{\epsilon_c}(x_c) \text{ or } j \in \mathcal{B}^{\epsilon_c}(x_c), \\ (H_c)_{ij} & \text{otherwise.} \end{cases} \tag{3.39}$$

When the explicit dependence on x_c, ϵ_c, and H_c is important we will write

$$\mathcal{R}(x_c, \epsilon_c, H_c).$$

So, for example,

$$\nabla_R^2 f(x_c) = \mathcal{R}(x_c, 0, \nabla^2 f(x_c)).$$

Given $0 < \epsilon < \min(U_i - L_i)/2$ and an spd H, define

$$x^{H,\epsilon}(\lambda) = \mathcal{P}(x - \lambda \mathcal{R}(x, \epsilon, H)^{-1}\nabla f(x)).$$

The appropriate sufficient decrease condition could be

$$f(x^{H,\epsilon}(\lambda)) - f(x) \leq -\frac{\alpha}{\lambda}\|x - x^{H,\epsilon}(\lambda)\|^2, \tag{3.40}$$

which becomes (3.33) if $H = I$, or

$$f(x^{H,\epsilon}(\lambda)) - f(x) \leq -\alpha\nabla f(x)^T(x - x^{H,\epsilon}(\lambda)), \tag{3.41}$$

which becomes (3.34) when $H = I$.

An algorithm based on these ideas is the **scaled gradient projection** [12] algorithm. The name comes from the **scaling matrix** (an archaic name for the model Hessian) H that is used to compute the direction. Left unstated in the algorithmic description are the manner in which the parameter ϵ is computed and the way in which the approximate Hessians are constructed. Note that the gradient projection method is a special case ($H = I$) of the scaled gradient projection method.

Algorithm 3.4.
sgradproj$(x, f, \tau, kmax)$

 $k = 1$.
 Compute f and ∇f; terminate if (3.36) holds.
 while $k \leq kmax$ and $\|x - x(1)\| > \tau$ **do**
  Compute ϵ and an H_c so that $\mathcal{R}(x, \epsilon, H_c)$ is spd.
  Solve

$$\mathcal{R}(x, \epsilon, H_c)d = -\nabla f(x_c).$$

  Find the least integer m such that (3.40) holds for $\lambda = \beta^m$.
  $x \leftarrow x^{H, \epsilon}(\lambda)$.
  $k \leftarrow k + 1$.
  Compute f and ∇f
 end while
 if $\|x - x(1)\| < \tau$ **then**
  signal success
 else
  signal failure
 end if

It is not important that the model Hessian H be spd, only that $\mathcal{R}(x, \epsilon, H)$ is spd. This is why it can be a problem to overestimate the nonbinding set. Underestimating the nonbinding set will, at worst, replace columns and rows of H that may cause indefiniteness with columns and rows of the identity.

Theorem 3.14. *Let ∇f be Lipschitz continuous with Lipschitz constant L. Assume that the matrices H_n are spd and the sequences $\{\|H_k\|\}$ and $\{\|H_k^{-1}\|\}$ are bounded. Assume that there is $\bar{\epsilon} > 0$ such that $\bar{\epsilon} \leq \epsilon_n < \min(U_i - L_i)/2$ for all n.*
 Then

$$\lim_{n \to \infty} \|x_n - x_n(1)\| = 0, \tag{3.42}$$

and hence any limit point of the sequence of iterates produced by Algorithm sgradproj *is a stationary point.*
 In particular, if $x_{n_l} \to x^*$ is any convergent subsequence of $\{x_n\}$, then $x^* = x^*(1)$. If x_n converges to a local minimizer x^* which satisfies the second-order sufficiency conditions, then the binding set of x_n is the same as that of x^* after finitely many iterations.

3.6 Fast Local Convergence

In general, the steepest descent and gradient projection methods converge slowly. By this we mean that it can take many iterations to reduce the norm of the gradient or projected gradient to the target level. This slow convergence is a particular problem when the iteration is near a solution.

 In the special case of **good data**, i.e., an initial iterate near to x^*, and an unconstrained problem, we can do much better by using the fact that the Hessian

$\nabla^2 f(x)$ will be spd if the second-order sufficiency conditions hold. This means that we can apply Newton's method for nonlinear equations [77, 106] and be certain that we will converge to x^* rather than a local maximum. Note that Newton's method is simply Algorithm **steeph** with $H_c = \nabla^2 f(x_c)$.

The bound constrained case is similar. If x_0 is near x^*, then we can use the scaled projected gradient method with $H_c = \nabla^2 f(x_c)$. This **projected Newton method** will converge as rapidly as Newton's method for nonlinear equations and identify the binding set in finitely many iterations as well.

We will state two convergence theorems, Theorem 3.15 for unconstrained problems and Theorem 3.16 for bound constrained problems. These results expose a central problem in continuous optimization. Near a solution, the Hessian is an almost perfect scaling matrix, and the iteration will converge rapidly. However, when far from an optimal point, the Hessian (or the modified Hessian) may not be spd and will be a useless scaling matrix unless something is done to manage negative curvature. In fact, when far from a solution, steepest descent (or gradient projection) will make very good progress. The challenge is to design a method that will behave like steepest descent (gradient projection) when far from a solution and then change into a fast local method when a solution is close. Of the several ways to meet this challenge, we have chosen the quasi-Newton methods, and we discuss those in § 3.8.

Theorem 3.15.   *Let f be twice Lipschitz continuously differentiable. Let x^* be a local minimum of f at which the second-order sufficiency conditions hold. Then if x_0 is sufficiently near x^*, the Newton iteration (Algorithm **steeph** with $H_c = \nabla^2 f(x_c)$) will take full steps ($\lambda = 1$ always) and converge **q-quadratically** to x^*, i.e.,*

$$\|x_+ - x^*\| = O(\|x_c - x^*\|^2).$$

Theorem 3.16.   *Let f be twice Lipschitz continuously differentiable. Let x^* satisfy the second-order sufficiency conditions. Then if x_0 is sufficiently near to x^* and $\mathcal{B}(x_0) = \mathcal{B}(x^*)$, the projected Newton iteration (Algorithm **sgradproj** with $H_c = \nabla^2 f(x_c)$), with $\epsilon_n = \|x_n - x_n(1)\|$, will converge q-quadratically to x^*.*

The assumptions of Theorems 3.15 and 3.16 are called the **standard assumptions** for convergence of Newton's method.

3.7   Finite Difference Approximations

Implicit filtering, as originally implemented in [75, 123, 123, 131], uses finite difference approximations for gradients and quasi-Newton models for the Hessian. Finite difference approximations have limits, and it's important to understand those limits. Assume for now that f is a real-valued function of a single real variable. The forward difference approximation

$$D_h^F f(x) = \frac{f(x+h) - f(x)}{h}$$

is accurate to first order if f is sufficiently smooth. Taylor's theorem implies that

$$D_h^F f(x) = f'(x) + \frac{f''(\xi)h}{2}, \tag{3.43}$$

where $\xi \in [x, x + h]$. The approximation will be acceptable for purposes of optimization if f'' is not too large. Similarly, the central difference approximation

$$D_h^C f(x) = \frac{f(x + h) - f(x - h)}{2h}$$

is second-order accurate, and

$$D_h^C f(x) = f'(x) + \frac{f'''(\xi)h^2}{6}, \tag{3.44}$$

and will be more accurate than the forward difference if h is small and f''' is not too large. One may assume less smoothness and still get the estimates like those in (3.43) and (3.44). For example, if f' is Lipschitz continuous with Lipschitz constant L, then

$$|D_h^F f(x) - f'(x)| \le \frac{Lh}{2}, \tag{3.45}$$

which we prove below. Similarly, if f'' is Lipschitz continuous with Lipschitz constant γ, we obtain

$$|D_h^C f(x) - f'(x)| \le \frac{\gamma h^2}{6}. \tag{3.46}$$

If one has very accurate derivative information, say, an analytic derivative, then it may be useful to approximate the second derivative with a difference. For the problems considered here, however, the function itself has limited accuracy, and approximation of a second derivative with a difference is unlikely to give good results.

Poor accuracy in f affects the quality of the approximation of the gradient because the formulae for difference approximations divide f by h. If we make h too small, we will be attempting to differentiate the errors in f. To see this, assume that we wish to differentiate a smooth function f_s but can only compute

$$f(x) = f_s(x) + \phi(x),$$

where ϕ is the error in f. We will refer to ϕ as the **noise** in f. A forward approximation will give

$$D_h^F f(x) = \frac{f(x + h) - f(x)}{h} + \frac{\phi(x + h) - \phi(x)}{h}.$$

Now, there is no reason to expect the term with the noise to be any smaller that $\|\phi\|_\infty/h$, so we write

$$D_h^F f(x) = f_s'(x) + O(h + \|\phi\|_\infty/h).$$

The quantity in the O-term is minimized when $h = \sqrt{\|\phi\|_\infty}$, and the error in the difference approximation of f'_s is $O(h)$. The forward difference approximation error is $O(1)$, i.e., hopeless, if $h \leq \|\phi\|_\infty$.

You may be familiar with the case in which ϕ is floating point roundoff ϵ_{mach}. In that case, the accuracy in the derivative is $O(\sqrt{\epsilon_{mach}})$. If we the use the forward difference derivative to compute a forward difference second derivative, we would get an error of $O(\epsilon_{mach}^{.25})$. The noise in many problems is far larger than floating point roundoff, and that is why numerical second derivatives are a problem. If f is smooth, then interpolation can be used to construct high-order models [36, 37, 38, 39, 112] if the points at which the interpolation is done are chosen carefully.

Centered differences improve this situation a bit. The error becomes

$$D_h^C f(x) = f'_s(x) + O(h^2 + \|\phi\|_\infty/h),$$

and the quantity inside the O-term is minimized when

$$h = (\|\phi\|_\infty/2)^{1/3}$$

and the error in the formula is $O(h^2)$, which is better than the error for forward differences. The approximation becomes useless if $h \leq \|\phi\|_\infty$, as was the case for forward differences.

If we knew the size of the noise, could predict its decay, or control it via changing the way we compute f, then we could manage the finite differences within the computation and control the accuracy. This has been done for implicit filtering [75] and other sampling methods [75, 88] as well as for traditional Newton-like methods [24, 25, 99]. We will discuss this in more detail when we talk about convergence rates in § 5.5.

Now we let $F : R^N \to R^M$ be a function of several variables. We assume that F' is Lipschitz continuous with Lipschitz constant γ. Let $x, v \in R^N$. The **directional derivative** of F at x in the direction v is

$$\partial F(x)/\partial v = F'(x)v.$$

The forward difference approximation of $\partial F/\partial v$ is

$$\frac{F(x + hv) - F(x)}{h}.$$

We will evaluate the accuracy of this approximation with Theorem 3.2. We begin with

$$F(x+hv)-F(x) = \int_0^1 F'(x+thv)^T hv\, dt = hF'(x)v + h\int_0^1 (F'(x+thv)-F'(x))^T v\, dt$$

and so

$$\left\| \frac{F(x+hv)-F(x)}{h} - F'(x)v \right\| \leq \int_0^1 \|F'(x+thv) - F'(x)\|\, dt \|v\|$$

$$\leq \|v\| \int_0^1 \gamma th\|v\|\, dt = \frac{\gamma h\|v\|^2}{2}.$$

(3.47)

If F is twice Lipschitz continuously differentiable, then the central difference is second-order accurate,

$$\left\| \frac{F(x+hv) - F(x-hv)}{2h} - F'(x)v \right\| = O(h^2\|v\|^2). \tag{3.48}$$

Implicit filtering uses finite differences and evaluations of F to approximate Jacobians and gradients. To illustrate how this goes, we will express a finite difference Jacobian as a solution of a linear least squares problem. Suppose V_1 is an orthogonal $N \times N$ matrix with columns $\{v_j^1\}$. Let $\delta(F, x, V_1, h)$ be the $M \times N$ matrix having $F(x + hv_j^1) - F(x)$ as its jth column. Taylor's theorem says that

$$\delta(F, x, V_1, h) = hF'(x)V_1 + O(h^2). \tag{3.49}$$

We can use (3.49) to recover a first-order accurate approximation to F' and define

$$DF(x, V_1, h) = \frac{1}{h}\delta(F, x, V_1, h)V_1^T = F'(x) + O(h). \tag{3.50}$$

We can express a centered difference approximation in a similar way. Since

$$\delta(F, x, V_1, h) - \delta(F, x, -V_1, h) = 2hF'(x)V_1 + O(h^3),$$

a second-order accurate approximation to F' is

$$DF(x, V_2, h) = \frac{1}{2h}\delta(F, x, V_2, h)V_2^T = F'(x) + O(h^2), \tag{3.51}$$

where $V_2 = [V_1, -V_1]$.

More generally, let V be any $N \times K$ matrix whose columns may be neither orthogonal, of unit length, or even linearly independent. We seek an approximate Jacobian $DF(x, V, h)$ which minimizes

$$\|\delta(F, x, V, h) - hDF(x, V, h)V\|.$$

If we let $V^\dagger$ be the pseudoinverse of V as defined by (3.10), then the solution is the **stencil derivative** or **stencil Jacobian**

$$DF(x, V, h) = \frac{1}{h}\delta(F, x, V, h)V^\dagger. \tag{3.52}$$

In the case where $V = V_1$ is an orthogonal matrix, then $V^\dagger = V^T$ and we recover the one-sided difference from (3.50). If $V = V_2 = [V_1, -V_1]$, then we recover the central difference formula (3.51). To see this, we first compute the SVD of V_2,

$$V_2 = \tilde{U}_2 \tilde{\Sigma}_2 \tilde{V}_2^T.$$

Since there are more columns than rows, we have

$$\tilde{\Sigma}_2^2 = V_2 V_2^T = 2I_{N \times N},$$

where $I_{N \times N}$ is the $N \times N$ identity matrix. Since the columns of $\tilde{V}_2$ are the eigenvectors of $V_2 V_2^T$,

$$\tilde{V}_2 = I_{N \times N}.$$

Having computed $\tilde{\Sigma}_2$ and $\tilde{V}_2$, it is clear that

$$\tilde{U}_2 = \frac{1}{\sqrt{2}} V_2.$$

Hence

$$V_2^T / 2 = V_2^\dagger,$$

and hence (3.52) and (3.51) agree. The most simple case is if we use the coordinate directions. In that case $V = I$ or $V = [I, -I]$.

3.8 Quasi-Newton Methods

One way to manufacture useful model Hessians is to approximate the Hessian in a way that reflects the history of the iteration and (sometimes) ensures that the model Hessian is positive definite. The idea is to find a sequence of model Hessians that perform well when the iteration is far from a local minimizer and converge rapidly when near a point which satisfies the second-order sufficiency conditions.

In this context, rapid convergence means **q-superlinear** convergence, i.e., either $x_n = x^*$ for all n sufficiently large or

$$\lim_{n \to \infty} \frac{\|x_{n+1} - x^*\|}{\|x_n - x^*\|} = 0. \tag{3.53}$$

Quasi-Newton methods update an approximation of $\nabla^2 f(x^*)$ as the iteration progresses. In the unconstrained case the transition from current approximations x_c and H_c of x^* and $\nabla^2 f(x^*)$ to new approximations x_+ and H_+ follows this paradigm:

1. Compute a search direction $d = -H_c^{-1} \nabla f(x_c)$.

2. Find $x_+ = x_c + \lambda d$ using a line search to ensure sufficient decrease.

3. Use x_c, x_+, and H_c to **update** H_c and obtain H_+.

The way in which H_+ is computed determines the method.

Quasi-Newton methods are also called **secant methods** because they satisfy the **secant equation**

$$H_+ s = y. \tag{3.54}$$

In (3.54)

$$s = x_+ - x_c \quad \text{and} \quad y = \nabla f(x_+) - \nabla f(x_c).$$

If $N = 1$, all secant methods reduce to the classical **secant method** for the single nonlinear equation $f'(x) = 0$, i.e.,

$$x_+ = x_c - \frac{f'(x_c)(x_c - x_-)}{f'(x_c) - f'(x_-)}. \tag{3.55}$$

We use two quasi-Newton methods in **imfil.m**. The BFGS (Broyden–Fletcher–Goldfarb–Shanno) [18, 50, 59, 119] method is supported by the most complete theory. The BFGS update preserves positive definiteness, and so the convergence theory for Algorithm **steeph** is, in principal, applicable.

The BFGS update is the rank-two update

$$H_+ = H_c + \frac{yy^T}{y^T s} - \frac{(H_c s)(H_c s)^T}{s^T H_c s}. \tag{3.56}$$

The **symmetric rank-one (SR1)** update is

$$H_+ = H_c + \frac{(y - H_c s)(y - H_c s)^T}{(y - H_c s)^T s}. \tag{3.57}$$

The SR1 update has been reported to out-perform BFGS algorithms in certain cases [21, 32, 33, 57, 80, 81, 84, 123, 125, 130], in which the approximate Hessians can be expected to be positive definite, the problem has bound constraints, or a trust-region framework is used. While there is some limited convergence theory for the SR1 update [21, 84], that theory does not apply to the way the method is used in **imfil.m**. The SR1 update is an option in **imfil.m**.

The default in **imfil.m** is the BFGS update.

3.8.1   Unconstrained Problems

The local theory for BFGS is completely satisfactory. When the data (x_0 and H_0) are good, the iteration will converge rapidly. Moreover, one need not compute or use a difference to approximate $\nabla^2 f$, which can be a significant advantage.

Theorem 3.17.   *Let f be twice Lipschitz continuously differentiable. Let x^* be a local minimum of f at which the second-order sufficiency conditions hold. Then if x_0 is sufficiently near x^* and H_0 sufficiently near $\nabla^2 f(x^*)$, the BFGS iteration (Algorithm steeph with H given by (3.56)) exists (i.e., $y_k^T x_k > 0$ for all k), will take full steps ($\lambda = 1$ always), and converges q-superlinearly to x^*.*

We can apply Theorem 3.11 if the BFGS updates $\{H_k\}$ remain bounded and well-conditioned. Assuming this, we would have a complete story if we could prove that if the iteration converged to a local minimizer x^* which satisfied the second-order sufficiency conditions, then the iteration was q-superlinearly convergent in the terminal phase (i.e., once the iteration got close enough to x^* for Theorem 3.17). The global convergence theory for BFGS is almost what we need but requires an assumption on the level sets of f that is very hard to verify. Having said that, Theorem 3.18 well describes what we often see in practice. One should think of Theorem 3.18 [22, 23] as describing what happens after the BFGS iteration has found a point near x^*, a local minimizer which satisfies the second-order necessary conditions. So, we have an accurate initial iterate but should expect the Hessian approximation to be poor.

Theorem 3.18. *Let f be twice Lipschitz continuously differentiable. Let x^* be a local minimizer of f which satisfies the second-order sufficiency conditions. Assume that H_0 is spd and that the set*

$$D = \{x \mid f(x) \le f(x_0)\}$$

is convex. Then if x_0 is sufficiently near x^*, the BFGS–Armijo iteration converges q-superlinearly to x^*

Theorems 3.17 and 3.18 say that we can expect the BFGS iteration to resolve the slow convergence problems of steepest descent, provided the BFGS model Hessians remain bounded and well-conditioned. Most implementations, **imfil.m** included, monitor $y^T x$ and skip the update if $y^T x$ is not sufficiently positive. This simple safeguard is normally enough to ensure good performance in practice.

3.8.2 Projected Quasi-Newton Methods

It may be the case that one can compute part of the model Hessian exactly. This will certainly be the case with bound constrained problems. We must manage the update in a way so that

- the computed part is not destroyed when the update is applied, and

- the secant equation holds.

We describe a simple artifice [44, 46, 75] for doing this.
Suppose

$$\nabla^2 f(x) \approx H = C(x) + A,$$

where $C(x)$ is the part of the Hessian one can compute exactly and cheaply. The quasi-Newton method will exploit this structure and only update A, the approximated part. So

$$H_+ = C(x_+) + A_+.$$

Superlinear convergence proofs require, in one way or another, that $H_+ s = y$. Therefore, in terms of A, one might require the update to satisfy

$$A_+ s = y^\# = y - C(x_+)s. \tag{3.58}$$

This is called the **default choice** in [46]. The secant equation $H_+ s = y$ will follow from (3.58). While there are other choices for $y^\#$ [44, 46], (3.58) is by far the most common and is what we use throughout.

Updates of this type will converge superlinearly if the underlying quasi-Newton update does. Here is a local convergence result [46] for the BFGS formulation for unconstrained problems:

$$A_+ = A_c + \frac{y^\# y^{\#T}}{y^{\#T} s} - \frac{(A_c s)(A_c s)^T}{s^T A_c s}. \tag{3.59}$$

Theorem 3.19. *Let f be twice Lipschitz continuously differentiable and let x^* be a local minimizer of f which satisfies the second-order sufficiency conditions. Assume that*

$$A^* = \nabla^2 f(x^*) - C(x^*)$$

is spd. Then there is δ such that if

$$\|x_0 - x^*\| \le \delta \text{ and } \|A_0 - A^*\| \le \delta,$$

then the quasi-Newton iterates defined by (3.59) exist and converge q-superlinearly to x^*.

We apply this to bound constrained optimization with

$$C(x) = \mathcal{P}_{\mathcal{B}^\epsilon(x)} \tag{3.60}$$

and update an approximation to the part of the model Hessian that acts on the ϵ-nonbinding set. In this way we can hope to maintain a positive definite model modified Hessian with, say, a BFGS update. So if our model modified Hessian is

$$\mathcal{R} = C(x) + A,$$

we can use (3.59) to update A (with $A_0 = \mathcal{P}_{\mathcal{N}^{\epsilon_0}(x_0)}$, for example), as long as the ϵ-nonbinding set does not change. If one begins the iteration near a nondegenerate local minimizer with an accurate approximation to the Hessian, then one would expect, based on Theorem 3.16, that the active set would remain constant and that the iteration would converge q-superlinearly by Theorem 3.19.

If the initial data is far from a local minimizer, the active set can change with each iteration and the update must be designed to account for this. One way to do this [75] is to use a projected form of the BFGS update of A from (3.59),

$$A_+ = \mathcal{P}_{\mathcal{N}_+} A_c \mathcal{P}_{\mathcal{N}_+} + \frac{y^\# y^{\#T}}{y^{\#T} s} - \mathcal{P}_{\mathcal{N}_+} \frac{(A_c s)(A_c s)^T}{s^T A_c s} \mathcal{P}_{\mathcal{N}_+}, \tag{3.61}$$

with

$$y^\# = \mathcal{P}_{\mathcal{N}_+}(\nabla f(x_+) - \nabla f(x_c)).$$

Here $\mathcal{N}_+ = \mathcal{N}^{\epsilon_+}(x_+)$. This update carries as much information as possible from the previous model modified Hessian while taking care about proper approximation of the binding set. As in the unconstrained case, if $y^{\#T} s \le 0$ (or less than some small positive number), we can either skip the update or reinitialize A to $\mathcal{P}_{\mathcal{N}}$. If one takes the latter approach, then once the active set has been identified, one has the same dilemma as in the unconstrained case, namely, an accurate x and a possibly inaccurate modified Hessian.

Similarly, the projected SR1 update is

$$A_+ = \mathcal{P}_{\mathcal{N}_+} A_c \mathcal{P}_{\mathcal{N}_+} + \mathcal{P}_{\mathcal{N}_+} \frac{(y^\# - A_c s)(y^\# - A_c s)^T}{(y^\# - A_c s)^T s} \mathcal{P}_{\mathcal{N}_+}. \tag{3.62}$$

For completeness, we present the convergence results for the projected BFGS iteration.

Theorem 3.20. *Let f be twice continuously Lipschitz continuously differentiable. Let x^* be a nondegenerate local minimizer which satisfies the second-order sufficiency conditions. Then if x_0 is sufficiently near to x^*, $\mathcal{B}(x_0) = \mathcal{B}(x^*)$, and A_0 is sufficiently near to $\mathcal{P}_{\mathcal{N}(x^*)} \nabla^2 f(x^*) \mathcal{P}_{\mathcal{N}(x^*)}$, then the projected BFGS iteration, with $\epsilon_n = \|x_n - x_n(1)\|$, will converge q-superlinearly to x^*.*

Theorem 3.21. *Let ∇f be Lipschitz continuous on Ω. Assume that the matrices H_n are constructed with the projected BFGS method (3.61) and satisfy the assumptions of Theorem 3.14. Then (3.42) and the conclusions of Theorem 3.14 hold.*

Moreover, if x^* is a minimizer satisfying the second-order sufficiency conditions, there is n_0 such that $\mathcal{B}(x_n) = \mathcal{B}(x^*)$ for all $n \geq n_0$, H_{n_0} is spd, and the set

$$D = \{x \,|\, f(x) \leq f(x_{n_0}) \text{ and } \mathcal{B}(x) = \mathcal{B}(x^*)\}$$

is convex, then the projected BFGS–Armijo iteration, with $\epsilon_n = \|x_n - x_n(1)\|$, converges q-superlinearly to x^*.

3.9 Nonlinear Least Squares

Nonlinear least squares problems have special structure which, as we saw in § 2.5.1, can be exploited to improve the performance of the iteration. We will describe that structure and show how the Gauss–Newton iteration exploits it.

Recall that the problem is

$$\min_{x} F(x)^T F(x)/2, \tag{3.63}$$

where $F : R^N \to R^M$ with $M > N$. It is easy to see [44, 75] that if

$$f(x) = F(x)^T F(x)/2,$$

then

$$\nabla f(x) = F'(x)^T F(x).$$

3.9.1 Gauss–Newton Iteration

One rarely computes $\nabla^2 f$ in practice because one has to compute the tensor F'' to do that. The algorithms we consider here use the **Gauss–Newton** model Hessian $F'(x)^T F'(x)$. The **Gauss–Newton iteration** from a current point x_c is

$$x_+ = x_c - (F'(x_c)^T F'(x_c))^{-1} F'(x_c)^T F(x_c). \tag{3.64}$$

The Gauss–Newton step

$$s = -(F'(x_c)^T F'(x_c))^{-1} F'(x_c)^T F(x_c)$$

is the solution of the linear least squares model problem

$$\min_x \|F'(x_c)s + F(x_c)\|$$

and can be solved (see § 3.1.3) with a QR factorization of F' to obtain

$$s = -R^{-1}Q^T F(x_c).$$

The local convergence for the Gauss–Newton iteration is different from the ones we've presented before in that there are restrictions on the size of $\|F\|$ at the solution. We begin by stating the **standard assumptions** for convergence of the Gauss–Newton method.

Assumption 1.

1. x^* *is a local minimizer of* $\|F(x)\|$.

2. F' *is Lipschitz continuous with Lipschitz constant* γ.

3. $F'(x^*)$ *has full column rank.*

Theorem 3.22. *Let* $F : R^N \to R^M$ *be Lipschitz continuously differentiable. Assume that Assumption 1 holds. Let* $x_c \in R^N$ *and let* x_+ *be given by (3.64). Then*

$$\|x_+ - x^*\| = O(\|x_c - x^*\|^2 + \|x_c - x^*\|\|F(x^*)\|). \tag{3.65}$$

Assumption 1 is not sufficient on its own for local convergence. As one can see from (3.65), not only must x_c be near x^* for x_+ to be a better approximation to x^*, but $\|F(x^*)\|$ must be small. Nonlinear least squares problems for which $\|F(x^*)\|$ is small enough for Gauss–Newton to converge are called **small residual problems**. This definition of small is, of course, imprecise. In the case of a **zero residual problem** where $F(x^*) = 0$, the convergence result is similar to Newton's method. The important point is that the local convergence is fast if $\|F(x^*)\|$ is small and that the Gauss–Newton iteration may diverge if $\|F(x^*)\|$ is too large. **imfil.m** is designed for small residual problems.

For the small, but nonzero, residual case, the Gauss–Newton iteration converges **q-linearly**, i.e., there is $\sigma \in (0,1)$ such that

$$\|x_+ - x^*\| \le \sigma \|x_c - x^*\| \tag{3.66}$$

whenever x_c is sufficiently near x^*. σ is called the **q-factor**. The convergence is called **r-linear** if

$$\|x_n - x^*\| \le Cr^n \tag{3.67}$$

for some $C > 0$ and $r \in (0,1)$. r is called the **r-factor**. This is weaker that q-linear convergence (3.66) in that there is no guarantee that the error norms are monotonically decreasing.

3.9.2 Projected Gauss–Newton Iteration

The **projected Gauss–Newton** method uses the Gauss–Newton model Hessian in the role of the Hessian in a projected Newton iteration. However, as in the unconstrained case, we avoid formation of $F'(x)^T F'(x)$ by using a QR factorization.

We begin with a simple version of the algorithm. The model Hessian at x_c is $H_c = F'(x_c)^T F(x_c)$. Using the notation of § 3.5.2 we let $\mathcal{B}^\epsilon(x_c)$ be the ϵ_c-binding set at x_c, $\mathcal{N}^{\epsilon_c}(x_c)$ its complement, and $\mathcal{P}_{\mathcal{N}^{\epsilon_c}}(x_c)$ and $\mathcal{P}_{\mathcal{B}^{\epsilon_c}}(x_c)$ the associated projections. The model of the modified Hessian is

$$\mathcal{R}_c^{GN} = \mathcal{P}_{\mathcal{B}^{\epsilon_c}}(x_c) + \mathcal{P}_{\mathcal{N}^{\epsilon_c}}(x_c)H_c\mathcal{P}_{\mathcal{N}^{\epsilon_c}}(x_c). \tag{3.68}$$

The equation for the projected Gauss–Newton direction is

$$\mathcal{R}_c^{GN}d = -F'(x_c)^T F(x_c), \tag{3.69}$$

so

$$\mathcal{P}_{\mathcal{B}^{\epsilon_c}}(x_c)d = -\mathcal{P}_{\mathcal{B}^{\epsilon_c}}(x_c)F'(x_c)^T F(x_c) \tag{3.70}$$

and

$$\mathcal{P}_{\mathcal{N}^{\epsilon_c}}(x_c)F'(x_c)^T F'(x_c)\mathcal{P}_{\mathcal{N}^{\epsilon_c}}(x_c)d = -\mathcal{P}_{\mathcal{N}^{\epsilon_c}}(x_c)F'(x_c)^T F(x_c). \tag{3.71}$$

Equation (3.70) is a direct formula for $\mathcal{P}_{\mathcal{B}^{\epsilon_c}}(x_c)d$ and (3.71) is nothing more than the normal equations for a linear least squares problem posed on the range of $\mathcal{P}_{\mathcal{N}^{\epsilon_c}}(x_c)$, namely,

$$\min_d \|F'(x_c)\mathcal{P}_{\mathcal{N}^{\epsilon_c}}(x_c)d + F(x_c)\|,$$

which we can solve with a QR factorization. If we let

$$F'(x_c)\mathcal{P}_{\mathcal{N}^{\epsilon_c}}(x_c) = QR,$$

with the understanding that we have deleted the zero columns in $F'(x_c)\mathcal{P}_{\mathcal{N}^{\epsilon_c}}(x_c)$, then

$$\mathcal{P}_{\mathcal{N}^{\epsilon_c}}(x_c)d = -R^{-1}Q^T F(x_c). \tag{3.72}$$

We will call $F'(x_c)\mathcal{P}_{\mathcal{N}^{\epsilon_c}}(x_c)$ the **reduced Jacobian** at x_c.

One can apply the Armijo rule to the Gauss–Newton iteration simply by using $\mathcal{R}^{GN}$ in Algorithm sgradproj. The resulting method is sometimes referred to as **damped Gauss–Newton**. Damped Gauss–Newton is the default nonlinear least squares solver in **imfil.m**.

The Gauss–Newton model Hessian is a good choice for small residual problems, and the convergence result [44, 75] for damped Gauss–Newton is strong. The convergence result we state for bound constrained problems follows from that for the unconstrained case [44, 75] and Theorem 3.14.

Theorem 3.23. *Let F be Lipschitz continuously differentiable on Ω and assume that $F'(x)$ has full column rank for all $x \in \Omega$. Let $\{x_n\}$ be the damped Gauss–Newton iterations. Then any limit point of $\{x_n\}$ is a stationary point. Moreover, if a limit point x^* is a local minimizer and $\|F(x^*)\|$ is sufficiently small, then $\lambda_n = 1$ for n sufficiently large, $x_n \to x^*$, and*

$$\|x_{n+1} - x^*\| = O(\|x_n - x^*\|^2 + \|F(x^*)\|\|x_n - x^*\|).$$

3.9.3   Levenberg–Marquardt Iteration

If F' becomes rank-deficient or nearly rank-deficient in the middle of a Gauss–Newton iteration, as it may do while far from optimality, then the Gauss–Newton direction may be a poor choice for a descent direction. In that event many step size reductions will be needed and the iteration may even stagnate. Theorem 3.23 must, therefore, make an assumption that F' has full column rank.

The **Levenberg–Marquardt** [90, 95] algorithm addresses this possibility of rank-deficiency by adding a regularization parameter ν to the Gauss–Newton model Hessian. The iteration is $x_+ = x_c + s$, where the trial step is

$$s = -(\nu_c I + F'(x_c)^T F'(x_c))^{-1} F'(x_c)^T F(x_c), \qquad (3.73)$$

where I is the $N \times N$ identity matrix. The matrix $\nu_c I + F'(x_c)^T F'(x_c)$ is positive definite if $\nu > 0$. The parameter ν is called the **Levenberg–Marquardt parameter**.

One can compute the Levenberg–Marquardt step [43] by solving the full-rank $(M + N) \times N$ linear least squares problem

$$\min \frac{1}{2} \left\| \begin{bmatrix} F'(x_c) \\ \sqrt{\nu_c} I \end{bmatrix} s + \begin{bmatrix} F(x_c) \\ 0 \end{bmatrix} \right\|^2. \qquad (3.74)$$

It is easy to see that the two approaches are the same in exact arithmetic, since the formula for s in (3.73) is the solution of the normal equations for (3.74).

3.9.4   Projected Levenberg–Marquardt Iteration

For bound constrained optimization, the Levenberg–Marquardt direction d is the solution of

$$(\mathcal{P}_{\mathcal{B}_c}(x_c) + \mathcal{P}_{\mathcal{N}_c}(x_c)(\nu_c I + F'(x_c)^T F'(x_c))\mathcal{P}_{\mathcal{N}_c}(x_c))d = -F'(x_c)^T F(x_c). \quad (3.75)$$

So in the binding directions we get

$$\mathcal{P}_{\mathcal{B}_c}(x_c)d = -\mathcal{P}_{\mathcal{B}_c}(x_c)F'(x_c)^T F(x_c),$$

which is the same as (3.70). In the nonbinding directions we get an analogue of (3.71),

$$\mathcal{P}_{\mathcal{N}_c}(x_c)(\nu I + F'(x_c)^T F'(x_c))\mathcal{P}_{\mathcal{N}_c}(x_c)d = -\mathcal{P}_{\mathcal{N}_c}(x_c)F'(x_c)^T F(x_c), \qquad (3.76)$$

which is the normal equations form of the following linear least squares problem on the range of $\mathcal{P}_{\mathcal{N}_c}(x_c)$,

$$\min \frac{1}{2} \left\| \begin{bmatrix} F'(p_c)\mathcal{P}_{\mathcal{N}_c}(x_c) \\ \sqrt{\nu_c}\mathcal{P}_{\mathcal{N}_c}(x_c) \end{bmatrix} d + \begin{bmatrix} F(x_c) \\ 0 \end{bmatrix} \right\|^2, \qquad (3.77)$$

which one can solve with an SVD or QR decomposition.

Finally, we explain a typical trust-region approach [35, 44, 66, 75] for managing ν. For simplicity we describe the unconstrained case. The Levenberg–Marquardt iteration x_+ is the minimum of the quadratic model

$$m_c(x) = f(x_c) + \nabla f(x_c)^T (x - x_c) + \frac{1}{2}(x - x_c)^T (\nu I + F'(x_c)^T F'(x_c))(x - x_c).$$

From a current point x_c and current value of the Levenberg–Marquardt parameter ν_c, one computes a trial point

$$x_t = \mathcal{P}(x_c + d), \tag{3.78}$$

where d is the solution of (3.77). The next step is to compare the actual reduction in f,

$$ared = f(x_c) - f(x_t),$$

to the reduction predicted by the quadratic model

$$pred = m_c(x_c) - m_c(x_t)$$

$$= -s^T \nabla f(x_c) - \tfrac{1}{2}s^T (\nu_c I + F'(x_c)^T F'(x_c))s,$$

where $s = x_t - x_c$.

As is the case with a line search method, one must decide whether to accept or reject the step. One also has to manage the Levenberg–Marquardt parameter. One common way to do this is to reject the step and increase ν if the ratio $ared/pred$ is too small, accept the step and decrease ν if $ared/pred$ is sufficiently large, and do something in between otherwise. In the formulation from [75] we also set the Levenberg–Marquardt parameter to zero when possible in order to recover the Gauss–Newton method's fast convergence for small residual problems. We do this by introducing yet another parameter ν_0, which we use as a lower bound for nonzero values of ν.

To quantify this we must introduce a number of parameters:

$$0 < \omega_{down} < 1 < \omega_{up}, \quad \nu_0 \geq 0, \quad \text{and} \quad 0 \leq \mu_0 < \mu_{low} \leq \mu_{high} < 1.$$

A typical choice is

$$\mu_0 = 0, \quad \mu_{low} = 1/4, \quad \mu_{high} = 3/4, \quad \omega_{down} = 1/2, \quad \text{and} \quad \omega_{up} = 2.$$

It's simplest to describe the method for managing ν formally as an algorithm.

Algorithm 3.5.
$\mathtt{Manage\_nu}(x_c, x_t, x_+, f, \nu)$

1. $z = x_c$.

2. Do while $z = x_c$

    (a) $ared = f(x_c) - f(x_t)$, $s = x_t - x_c$.

(b) $pred = -s^T \nabla f(x_c) - \frac{1}{2}s^T(\nu_c I + F'(x_c)^T F'(x_c))s$.

(c) If $ared/pred < \mu_0$, then set $z = x_c$, $\nu = \max(\omega_{up}\nu, \nu_0)$, and recompute the trial point with the new value of ν.

(d) If $\mu_0 \le ared/pred < \mu_{low}$, then set $z = x_t$ and $\nu = \max(\omega_{up}\nu, \nu_0)$.

(e) If $\mu_{low} \le ared/pred$, then set $z = x_t$.
    If $\mu_{high} < ared/pred$, then set $\nu = \omega_{down}\nu$.
    If $\nu < \nu_0$, then set $\nu = 0$.

3. $x_+ = z$.

The algorithm for controlling ν fits into the overall method in a manner similar to the way the line search enters into the Gauss–Newton iteration.

Algorithm 3.6.
LevMar$(x, F, \epsilon, kmax)$

1. Set $\nu = \nu_0$.

2. For $k = 1, \dots, kmax$

    (a) Let $x_c = x$.

    (b) Compute F, f, F', and ∇f; test for termination.

    (c) Compute x_t using (3.78).

    (d) Call Manage_nu(x_c, x_t, x, f, ν).

Theorem 3.24 [103, 133] is one convergence result.

Theorem 3.24. *Let F be Lipschitz continuously differentiable. Let $\{x_k\}$ and $\{\nu_k\}$ be the sequence of iterates and Levenberg–Marquardt parameters generated by Algorithm* levmar *with $kmax = \infty$. Assume that $\{\nu_k\}$ is bounded from above. Then either $F'(x_k)^T F(x_k) = 0$ for some finite k or*

$$\lim_{k \to \infty} F'(x_k)^T F(x_k) = 0.$$

Moreover, if x^* is a limit point of $\{x_k\}$ for which $F(x^*) = 0$ and $F'(x^*)$ has full rank, then $x_k \to x^*$ q-quadratically and $\nu_k = 0$ for k sufficiently large.

While the damped Gauss–Newton iteration is the default in **imfil.m**, you can replace it with Levenberg–Marquardt by using the **executive** option (see § 7.5 and § 8.4.2).

Part II

Algorithms and Theory

Chapter 4

The Implicit Filtering Algorithm

We begin this chapter on the algorithm by asking the reader to remember that implicit filtering is not designed to solve the classical smooth nonlinear programming problem and that there are much better choices for general-purpose nonlinear programming.

We think of implicit filtering as an extension of coordinate search, the simplest possible approach to optimization. Implicit filtering differs from coordinate search in that it builds a **local surrogate** for or **local model** of the objective function using a quasi-Newton method.

4.1 Coordinate Search

The classical coordinate search algorithm is for **unconstrained optimization**, i.e., problem (1.1) where $\Omega = R^N$. The algorithm begins with a **base point** x, the value $f(x)$, and a **scale** h. The algorithm begins by evaluating f evaluated at the $2N$ points on the **stencil**

$$S(x, h) = \{z \mid z = x \pm he_i\} \qquad (4.1)$$

centered at x. In (4.1) e_i is the unit vector in the ith coordinate direction. In a **nonopportunistic search** we sample the entire stencil and replace x with x_{min}, where

$$f(x_{min}) = \min_{z \in S(x,h)} f(z)$$

if $f(x_{min}) < f(x)$. If $f(x_{min}) \geq f(x)$, we **shrink** the stencil by reducing h, by a factor of 2, for example.

The final thing one needs in order to have a working method is a termination criterion. Gradient-based methods for smooth problems [44, 51, 56, 75, 108] can test for approximate satisfaction of the necessary conditions for optimality. One common way to control the cost of a sampling method is to give the iteration a **budget** of calls to f and to terminate the iteration when that budget is exhausted.

This simple algorithm works, albeit slowly, and is easy to understand both theoretically and practically. Algorithm `cs_basic` is the foundation of all the algorithms in this book. We will let V be the $N \times 2N$ matrix of positive and negative coordinate directions,

$$V = \{\pm e_i \,|\, 1 \leq i \leq N\}, \tag{4.2}$$

in our search algorithms for now.

Algorithm 4.1.
$(x_{opt}, f_{opt}, f_{count}) = \textbf{cs\_basic}(x, f, h, budget)$

    $f_{base} = f(x); \; f_{count} = 1;$
    $V = (e_1, e_2, \ldots, e_N, -e_1, \ldots, -e_N)$
    while $f_{count} \leq budget$ **do**
        for $i = 1, \ldots, 2N$ **do**
            $F_i^{out} = f(x + hv_i)$
        end for
        $f_{count} = f_{count} + 2N$
        $x_t = x + hv_i^*$ where $f_{min} = f(x + hv_i^*) = \min_i(F_i^{out})$.
        if $f_{min} \geq f_{base}$ **then**
            $h \leftarrow h/2$
        else
            $f_{base} = f_{min}; \; x = x_t$
        end if
    end while
    $x_{opt} = x; \; f_{opt} = f_{base}$

    Figure 4.1 tracks the first two iterations of coordinate search. The points on the initial stencil are labeled with circles. The function values are next to the points on the plot. As you can see, the point to the right of the initial point in the stencil about that point is better (a value of 4 is better than the value of 5 at the initial point). The stencil about the first iterate x_1 does not contain a better point, so we reduce the scale (to get the points labeled with crosses) by half. When we do that we find x_2 with a function value of 2, which is better. The search would proceed from x_2 with the smaller scale. Another thing you can see from the figure is that some points are members of more than one stencil, and a good implementation would keep track of the history of the search and avoid evaluating f at the same point more than once.

    We have designed the algorithm to return the best point, the value at that point, and the cost. As we develop implicit filtering we will point out that the algorithm could return much more, and **imfil.m** gives you the option to recover the entire history of the iteration if you want it. We want to make Algorithm cs_basic as simple as possible and do not put a lower bound on h. Such a lower bound is, however, important in practice.

    Algorithm cs_basic illustrates two properties that most sampling methods share. The first is that the iteration will probably terminate over budget, i.e., f_{count} will be greater than *budget* when the algorithm terminates. The reasons for this are that we do not enforce a lower bound on h and that the function counter

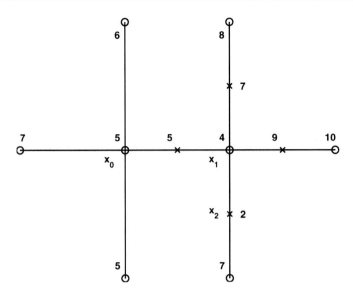

Figure 4.1. *Two iterations of coordinate search.*

is compared only to the budget at the top of the **while** loop. So

$$budget \leq f_{count} \leq budget + 2N - 1,$$

but $f_{count} = budget$ is very unlikely. We take the position that terminating the evaluation of f on the stencil midstream is a poor idea.

The second thing to notice is the opportunity to save some functions by exiting the inner for loop as soon as the search finds a better point. This **opportunistic search** would change the algorithm to the following.

Algorithm 4.2.
$(x_{opt}, f_{opt}, f_{count}) = \textbf{cs\_basic\_o}(x, f, h, budget)$
   $f_{base} = f(x)$; $f_{count} = 1$;
   while $f_{count} \leq budget$ **do**
      $f_{min} = f_{base}$
      while $f_{min} = f_{base}$ **do**
         if $f(x + hv_i) < f_{min}$ **then**
            $x \leftarrow x + hv_i$; $f_{min} = f(x)$;
         end if
         $f_{count} = f_{count} + 1$
      end while
      if $f_{min} = f_{base}$ **then**
         $h \leftarrow h/2$
      end if
   end while
   $x_{opt} = x$; $f_{opt} = f_{base}$

Algorithm cs_basic_o is in many ways more natural than the nonopportunistic version (think of an animal searching for food). Even so, implicit filtering is by design a nonopportunistic method. The reason for this choice is that implicit filtering uses the sampling to approximate a gradient and thereby find a direction for search and build a local model of f. Many other sampling methods are opportunistic [2, 68, 69, 88, 89].

When none of the function values on the stencil improve the base point, we say that a **stencil failure** [17] has happened. In Chapter 5 we will show how the use of stencil failure to control the size of the stencil enables us to prove convergence theorems. The analysis for the opportunistic methods also depends on stencil failure.

Several modern sampling algorithms may be thought of as elaborations on coordinate search. We will focus on the nonopportunistic algorithm and refer to [6, 88] for a more general algorithmic description, some of which we will describe in Chapter 5.

Implicit filtering does not require that the stencil consist of the positive and negative coordinate directions. One can base the search on any set of directions but must take some care in interpreting the results, as we will see in § 5.6. We will let

$$V = (v_1, \ldots, v_K) \tag{4.3}$$

be a matrix of directions and consider stencils which are based on the columns of V,

$$S(x, h, V) = \{z \mid z = x + hv_i, \ 1 \le i \le K\}. \tag{4.4}$$

The default stencil in **imfil.m** is the $2N$ point stencil based on the positive and negative coordinate directions. This is usually a good choice, but you may want to enlarge the stencil if there are more general constraints than simple bounds (see § 4.4.4 and § 5.6) or you may have to prune directions if f fails to return a value at $x + hv_i$ or $x + hv_i$ is infeasible (see § 4.4.3).

Algorithm cs_basic can be divided into a few parts, each of which will in turn be part of implicit filtering. The first and most critical thing Algorithm cs_basic does is to **poll** the points in the stencil. Our first version of a poll_stencil algorithm assumes that f always returns a value. As we will see later, that assumption is not always true, nor is it usually enough to just return the best point. In § 4.4.3 we show how the simple poll in this section must be extended. The simple algorithm **poll_stencil_v1** computes the values of f on a stencil $S(x, h, V)$, where V is an $N \times K$ matrix of directions. It returns a vector of function values, the best point on the stencil, and the function value at the best point.

Algorithm 4.3.
$(f_{min}, x_{min}, F^{out}) = $ **poll_stencil_v1**(x, f, h, V)
    for $i = 1, \ldots, K$ **do**
        $F_i^{out} = f(x + hv_i)$
    end for
    Find i^* so that $f(x + hv_{i^*}) = \min_i(F_i^{out})$.
    $x_{min} = x + hv_{i^*}$, $f_{min} = f(x_{min})$.

We will use and extend the poll step throughout the development of implicit filtering in this chapter. The role of polling the stencil in convergence (and the term "poll") was introduced in [16], but the ideas in that paper, which we use in some of the analysis in Chapter 5, have been in the literature at least since [68]. Algorithm test_stencil_v1 takes the output from poll_stencil_v1 and then tests for stencil failure, either updates x and $f(x)$ or reduces the scale h, and returns a flag that signals stencil failure or not. The flag will be important as we evolve coordinate search into implicit filtering.

Algorithm 4.4.
$(f_{base}, x_{base}, h, sflag) = \textbf{test\_stencil\_v1}(x_{base}, x_{min}, f_{base}, f_{min}, h)$

> **if** $f_{min} \geq f_{base}$ **then**
>> $h \leftarrow h/2$
>> $sflag = 1$
> **else**
>> $f_{base} = f_{min}; \; x_{base} = x_{min}$
>> $sflag = 0$
> **end if**

In our second version of coordinate search we use poll_stencil_v1 and have to specify our choice of search directions explicitly. We can use different sets of directions with equally good theoretical properties, and many sampling methods [7, 68, 69, 88] do that. We explain what useful direction sets look like in § 5.3.

Algorithm 4.5.
$(x_{opt}, f_{opt}, f_{count}) = \textbf{csearch}(x, f, h, budget)$

> $f_{base} = f(x); \; x_{base} = x; \; f_{count} = 1;$
> $V = (e_1, e_2, \ldots, e_N, -e_1, \ldots, -e_N)$
> **while** $f_{count} \leq budget$ **do**
>> $(f_{min}, x_{min}, F^{out}) = \textbf{poll\_stencil\_v1}(x, f, h, V)$
>> $f_{count} = f_{count} + 2N$
>> $(f_{base}, x_{base}, h, sflag) = \textbf{test\_stencil\_v1}(x_{base}, x_{min}, f_{base}, f_{min}, h)$
> **end while**
> $x_{opt} = x_{base}; \; f_{opt} = f_{base}$

4.2 Primitive Implicit Filtering

Implicit filtering extends coordinate search by approximating a gradient using the values of f on the stencil, uses that approximate gradient to build a model of f, and then searches for a better point using information from the model. One way to view this is that if the test of the stencil does find a better point (i.e., there is no stencil failure), then we try to use the function values we've computed to do even better.

4.2.1 The Stencil Gradient

Let V be a matrix of **directions**,

$$V = (v_1, \ldots, v_K),$$

where each column $v_i \in R^N$. In the case of central differences, $K = 2N$ and V is given by (4.2); that stencil is the default in **imfil.m**.

We will use the notation of § 3.7 to define the **stencil gradient** of f. Recall that the **stencil Jacobian** of a scalar-valued function f is the row vector

$$Df(x, V, h) = \frac{1}{h} \delta(f, x, V, h) V^{\dagger},$$

where

$$\delta(f, x, V, h) = \begin{pmatrix} f(x + hv_1) - f(x) \\ f(x + hv_2) - f(x) \\ \vdots \\ f(x + hv_K) - f(x) \end{pmatrix}^T.$$

We define the stencil gradient of a scalar-valued function f by

$$\nabla f(x, V, h) = Df(x, V, h)^T. \tag{4.5}$$

Hence the stencil gradient is the (minimum norm if $K < N$) solution of the linear least squares problem

$$\min_{y \in R^N} \| hV^T y - \delta(f, x, V, h)^T \|. \tag{4.6}$$

In [17, 75] we referred to the stencil gradient as a *simplex* gradient, as have others since then (see [38]). We prefer the name `stencil gradient` because the stencils we use here are more general that simplexes.

For nonlinear least squares problems, δF is an $N \times K$ matrix, $f = F^T F / 2$, and and we compute $\nabla f(x, V, h)$ by

$$\nabla f(x, V, h) = DF(x, V, h)^T F(x). \tag{4.7}$$

Recall from § 3.7 that if V is the matrix of central (forward) difference directions, one does indeed recover the usual centered (forward) difference gradient. The advantage of this formulation is that we can recover useful information if $K < N$ (as it might be if f is not defined at all points in the nominal design space) and give meaning to the stencil gradient for very general sets of directions.

imfil.m solves (4.6) with a QR factorization (see § 3.1.3) by computing a stencil Jacobian with (3.52). **imfil.m** computes the stencil Jacobian by differencing F. One might also approximate the derivatives of the components of F one at a time [135] and, perhaps, better deal with scaling, at a cost of algorithmic complexity.

4.2.2   Implementation of Primitive Implicit Filtering

This section describes the simplest possible form of implicit filtering, one for uncon-strained problems where f is everywhere defined. The essential idea is to augment coordinate search with a finite difference version of Algorithm **steep**, the method of steepest descent.

The logic in algorithm `imfil_basic` is more complex than either coordinate search or steepest descent. The algorithm begins with a sweep of coordinate search and a test for stencil failure. In the event of stencil failure, the scale is reduced and the steepest descent part of the while loop is not executed. If there is a better point on the stencil, then we compute the negative stencil gradient and use that as a search direction, terminating when

$$\|\nabla f(x, V, h)\| \le \tau h, \tag{4.8}$$

i.e., when the norm of the stencil gradient is the order of the stencil size. If (4.8) holds, we reduce the scale; otherwise we enter the line search. The scalar τ is a parameter in **imfil.m** and can be set with the `termtol` options. The default value is

$$\tau = \mathtt{imfil\_fscale} * \mathtt{termtol}.$$

Here `imfil_fscale` is a "typical value" of f and is set with the `fscale` option (see § 6.5.1). In **imfil.m** `imfil_fscale` is used to scale f to $O(1)$ at the beginning of the optimization. `termtol`, on the other hand, is the option designed to control τ (see § 6.11.4). The default choices for these options are

$$\mathtt{imfil\_fscale} = 1.2|f(x_0)| \quad \text{and} \quad \mathtt{termtol} = .01.$$

Unlike the line search for steepest descent, there is no guarantee, for a general h and smooth f, that the stencil gradient $-\nabla f(x, V, h)$ is a descent direction (see [40] for a more detailed discussion of this issue), so we must limit the number of step size reductions within the line search. `maxitarm` is that limit. It's a good idea to make that limit tight. In **imfil.m**, `maxitarm` = 3 is the default. If (3.25) fails to hold after `maxitarm` reductions in the step length, we say the **line search has failed**. However, the algorithm will begin the line search only if there is a better point in the stencil, so that better point will be a candidate for the next point, whether the line search succeeds or not. Hence (and unlike the early versions of implicit filtering [58]), failure of the line search will not cause a change in the scale on its own.

It's useful to describe the line search as a separate algorithm, because it will be used many times in the development. This version differs from the one in **imfil.m** only in that it is not aware of bound constraints.

We could extend (3.25) to use the stencil gradient and replace (3.29) with

$$f(x_c - \lambda d) - f(x_c) < -\alpha \lambda \nabla f(x, V, h)^T d. \tag{4.9}$$

Then the line search would terminate if (4.9) holds. In **imfil.m** we do not do this but only ask for a simple decrease:

$$f(x_c - \lambda d) < f(x_c). \tag{4.10}$$

One reason for this decision is that if $\nabla f(x, V, h)$ is a poor approximation of $\nabla f(x)$, (4.9) could be much more restrictive than simple decrease and thereby cause a good step to be rejected. Moreover, our convergence proof exploits stencil failure and does not need (4.9). This is one significant way that the analysis of sampling methods differs from that of gradient-based methods [17, 45, 126]; the proof of Theorem 3.11 very much depends on (3.29).

Algorithm 4.6.

$(f_{min}, x_{min}, acount) = \mathbf{Armijo\_linesearch\_v1}(f, x_c, d)$

    Find the least integer $m \geq 0$ such that (4.10) holds for $\lambda = \beta^m$.

    $x_{min} = x_c + \lambda d$

    $f_{min} = f(x_{min})$

    $acount = m + 1$

    So, we abandon a scale h and set $h \leftarrow h/2$ if we see either

- stencil failure or

- satisfaction of (4.8).

    We must also keep careful account of the number of calls of f. For each pass thorough the while loop we take $2N$ calls of f to evaluate f on the stencil. After that no more evaluations need to be done until we enter the line search. During the line search we must evaluate the function $m + 1$ times. If the line search succeeds, we will have already done the evaluation of $f_{base} = f(x)$ in the last line in the loop. So the total number of calls to f in a single pass of the while loop is at most $2N + m + 1$.

    The final detail is that we compare the value at the current point f_{base}, the value from the line search f_{newt}, and the best value from the stencil f_{min}. Clearly if $f_{newt} < f_{min}$, we accept the point from the line search x_{newt} as the new point. It is less clear what to do if

$$f_{min} < f_{newt} < f_{base}.$$

In the algorithmic sketch below, we adopt the default behavior in **imfil.m** and accept the quasi-Newton step if $f_{newt} < f_{base}$. We have found that **imfil.m** performs better if one favors the quasi-Newton step in this way. You can change this bias by setting the `stencil_wins` option to "yes" (see § 6.5.4).

Algorithm 4.7.

$\mathbf{imfil\_basic}(x, f, h, budget, h_{min}, maxitarm, \tau)$

    $f_{base} = f(x); \ f_{count} = 1; \ V = (e_1, e_2, \ldots, e_N, -e_1, \ldots, -e_N)$

    while $f_{count} \leq budget$ and $h \geq h_{min}$ **do**

        $z = x; \ f_{min} = f_{base}$

        $(f_{min}, x_{min}, F^{out}) = \mathbf{poll\_stencil}(x, f, h, V)$

        $f_{count} = f_{count} + 2N$

        $(f_{min}, x_{min}, h, sflag) = \mathbf{test\_stencil\_v1}(x_{base}, x_{min}, f_{base}, f_{min}, h)$

Compute $\nabla f(x, V, h)$.
if $sflag = 0$ and $\|\nabla f(x, V, h)\| > \tau h$ **then**
    if $\|\nabla f(x, V, h)\| \leq \tau h$ **then**
        $h \leftarrow h/2$
    else
        Set $d = -\nabla f(x, V, h)$.
        $(f_{newt}, x_{newt}, account) = \textbf{Armijo\_linesearch\_v1}(f, x, d)$.
        $fcount = fcount + acount$.
        if $f_{newt} < f_{base}$ **then**
            $x \leftarrow x_{newt}$ and $f_{base} = f_{newt}$.
        end if
    end if
end if
end while

Algorithm `imfil_basic` performs poorly in practice but is very close to a workable method. It's easy to use the ideas in § 3.5.2 to adapt `imfil_basic` to bound constrained problems and to add a model Hessian or some other scaling matrix to Algorithm **imfil_basic** by following the process through which Algorithm **steep** evolved to Algorithm **steeph**. The quasi-Newton or Gauss–Newton model Hessian is essential for good performance, and one might think that there's no more to do.

However, as we will see in § 4.4.3, that would be wrong [107].

4.3 Model Hessians and Bounds

4.3.1 Quasi-Newton and Gauss–Newton model Hessians

imfil.m uses a quasi-Newton (either BFGS or SR1) or Gauss–Newton model Hessian to accelerate convergence. The change in the algorithm is simply to use the quasi-Newton search direction

$$d = -H^{-1} f(x, V, h),$$

where H is the model Hessian, and then update the model Hessian with the stencil gradient after the line search is complete.

For nonlinear least squares problems **imfil.m** requires a function that returns the residual $F \in R^M$. **imfil.m** will then use that to compute the stencil gradient via (4.7) and will use the stencil Gauss–Newton

$$H^{GS}(F, x, V, h) = DF(x, V, h)^T DF(x, V, h) \tag{4.11}$$

or Levenberg–Marquardt (if you ask via the `executive_function` option)

$$\nu I + H^{GS}(F, x, V, h)$$

model Hessian. Gauss–Newton is the default, and the search direction is

$$d = -(DF(x, V, h)^T DF(x, V, h))^{-1} DF(x, V, h)^T F(x), \tag{4.12}$$

which is the solution of the linear least squares problem

$$\min \|DF(x, V, h)d + F(x)\|. \tag{4.13}$$

Similarly, the Levenberg–Marquardt step is

$$s = -(\nu I + DF(x, V, h)^T DF(x, V, h))^{-1} DF(x, V, h)^T F(x). \tag{4.14}$$

For the Gauss–Newton method, the changes to Algorithm imfil_basic are only in the computation of the search direction, where now $d = -H_c^{-1}\nabla f(x, V, h)$ and $H_c = DF(x, V, h)^T DF(x, V, h)$. In the case of a Levenberg–Marquardt step, one manages ν with the algorithm manage_nu from § 3.9.3.

4.3.2 Bound Constraints and Scaling

imfil.m is designed for bound constrained problems and expects the user (i.e., you) to provide sensible bounds for the variables. One reason for this is that good bounds can be used to **scale** the variables. By this we mean that we can change variables to make all the variables roughly the same size. In **imfil.m** we do that in two ways. First we scale f by a "typical value" to make the function values $O(1)$ during the optimization. The option fscale determines how we do that (see § 6.5.1). Second, we change variables so that the bounds are 0 and 1. Suppose, for example, that the given hyperrectangle is

$$\Omega = \{x \in R^N \mid L_i \le (x)_i \le U_i\}.$$

A difference Jacobian with a single increment h would be highly inaccurate if, say, some variables were between 1 and 2 and others between 10^6 and 10^7. **imfil.m** uses new variables z where

$$0 \le (z)_i = \frac{(x)_i - L_i}{U_i - L_i} \le 1.$$

imfil.m initializes the scales with $h = 1/2$.

The poll of the stencil will require that the points be in Ω. So in the bound constrained case

$$S(x, h) = \{z \mid z = x \pm he_i\} \cap \Omega.$$

Assuming that there is a better point in the stencil (i.e., *sflag* = 0) the **imfil.m** takes a finite difference projected quasi-Newton (or Gauss–Newton) iterate. **imfil.m** computes the model of the modified Hessian using the methods from § 3.8.2, § 3.9.2, or § 3.9.4. In the context of an implementation of implicit filtering, we do not expect to drive the scales to zero, so we can fix the parameter ϵ to a small value when we compute the set $\mathcal{B}^\epsilon$ in (3.37). We must also use the stencil derivatives rather than the partial derivatives, and this could make using the sets

$$\overline{\mathcal{B}}^\epsilon(x, V, h) = \{i \mid U_i - (x)_i \le \epsilon \text{ and } \nabla f(x, V, h)^T e_i \le -\sqrt{\epsilon}\}$$

$$\cup \{i \mid (x)_i - L_i \le \epsilon \text{ and } \nabla f(x, V, h)^T e_i \ge \sqrt{\epsilon}\} \tag{4.15}$$

a problem if h is not very small. So, while we use the sets $\overline{\mathcal{B}}^\epsilon(x, V, h)$ in the analysis in § 5.5.1 we use the larger sets

$$\mathcal{B}^\epsilon(x, V, h) = \{i \mid U_i - (x)_i \leq \epsilon\}$$

$$\cup \{i \mid (x)_i - L_i \leq \epsilon\}$$

(4.16)

in the implementation in order to avoid testing the sign of the difference approximations.

We will use the notation

$$\mathcal{B}^\epsilon \quad \text{or} \quad \mathcal{B}^\epsilon(x)$$

for $\mathcal{B}^\epsilon(x, V, h)$ when there will be no confusion.

In **imfil.m** $\epsilon = 10^{-6}$ is fixed for the entire optimization.

The line search will take as its input the quasi-Newton direction

$$d = -\mathcal{R}^{-1} \nabla f(x, V, h),$$

where $\mathcal{R}$ is defined similarly to (3.39),

$$\mathcal{R} = \mathcal{P}_{\mathcal{B}^\epsilon}(x) + (I - \mathcal{P}_{\mathcal{B}^\epsilon}(x)) H (I - \mathcal{P}_{\mathcal{B}^\epsilon}(x)).$$

4.4 The Implementation in imfil.m

Implicit filtering, as implemented in **imfil.m**, augments Algorithm **imfil_basic** with quasi-Newton methods, intelligent responses to constraint violations, more general stencils, and several ways to exploit parallelism.

An **outer iteration** simply keeps track of the scale h and the budget. For each h an **inner iteration** applies the appropriate method (quasi-Newton, Gauss–Newton). The inner iteration will check for stencil failure as it computes the stencil gradient or Jacobian and reduce h when it terminates.

We will describe the implementation in terms of a quasi-Newton minimization of a scalar function f. For nonlinear least squares problems the differences are that **imfil.m** uses the vector F rather than $f = F^T F / 2$ and computes a stencil Jacobian instead of a stencil gradient.

4.4.1 The Outer Iteration

Algorithm **imfil_outer** is a general description of the outer iteration and reflects the way **imfil.m** works. The inputs are an initial iterate x, the function f (which is scalar valued for general optimization problems and vector valued for nonlinear least squares), the initial value for h, the upper and lower bounds U and L, the stencil V, and the lengthy list of termination criteria $budget$, h_{min}, maxit, maxitarm, and τ.

Algorithm 4.8.

$\mathbf{imfil\_outer}(x, f, h, U, L, V, budget, h_{min}, maxit, maxitarm, \tau)$

    $f_{base} = f(x)$; $f_{count} = 1$;

    while $f_{count} \leq budget$ and $h \geq h_{min}$ **do**

        $(f_{base}, x, icount) = \mathbf{imfil\_inner}(x, f, U, L, V, h, maxit, maxitarm, \tau)$.

        $f_{count} = f_{count} + icount$; reduce h.

    end while

4.4.2   The Inner Iteration

Algorithm **imfil_inner** is an example of an inner iteration for the case of optimization with a quasi-Newton method. We have hidden many of the details in this sketch and will elaborate on them later in this chapter. For example, the poll step and the line search must be aware of the bound constraints, not sample infeasible points on the stencil, and be able to respond to failures in the function evaluation. We have already discussed the `maxitarm` parameter, which limits the number of step length reductions in the line search. The `maxit` parameter is the upper bound on the number of nonlinear (or inner) iterations.

Algorithm 4.9.

$(f_{base}, x, icount) = \mathbf{imfil\_inner}(x, f, U, L, V, h, maxit, maxitarm, \tau)$

    $p = 1$; $\epsilon = 10^{-6}$.

    Compute $f_{base} = f(x)$ and $\nabla f(x, V, h)$.

    while $p \leq pmax$ and $\|x - \mathcal{P}(x - \nabla f(x, V, h))\| \geq \tau h$ **do**

        Poll the stencil.

        if there is a better point in the stencil **then**

            record the best point x_{min} in the stencil and the value $f_{min} = f(x_{min})$.

        else

            terminate with success on stencil failure.

        end if

        Update the model Hessian H and compute $\mathcal{R} = \mathcal{P}_{\mathcal{B}^{c}}(x) + \mathcal{P}_{\mathcal{I}^{c}}(x)H\mathcal{P}_{\mathcal{I}^{c}}(x)$.

        Solve $\mathcal{R}d = -Df(x, V, h)f(x)$.

        Use a backtracking line search, with at most $maxitarm$ backtracks, to find a step length λ.

        if the line search succeeds and returns x_{newt} **then**

            $x \leftarrow x_{newt}$.

        else

            $x \leftarrow x_{min}$.

        end if

        $f_{base} = f(x)$ and $\nabla f(x, V, h)$.

        $p \leftarrow p + 1$

        Update $icount$.

    end while

    If $p > $ `maxit` report iteration count failure.

    If $p <= $ `maxit` report success.

In **imfil.m** you can limit the size of the quasi-Newton direction before beginning the line search. This is useful if you find that the line search fails many times. If you set the `limit_quasi_newton` (see § 6.5.5) option to "yes" (the default), the quasi-Newton direction will be limited to a length of $10h$. If your problem is smooth or nearly so, you should set `limit_quasi_newton` to "no." The case study in Chapter 8 is such a smooth problem if the tolerances for the integrator are set sufficiently tightly.

4.4.3 Hidden Constraints

We can use the poll step and the concept of stencil failure to handle constraints more general than simple bounds. While **imfil.m** can only handle bound constraints in a direct way, more general constraints can be sent to the poll step to avoid sampling infeasible points.

Implicit filtering was designed to deal with failures of the function to return a value [131]. We will refer to these failures as violations of **hidden constraints**. Hidden constraints are common and sampling algorithms must be prepared to deal with failures in the function. The idea of stencil failure enables one to deal both algorithmically and theoretically with these constraints. **imfil.m** expects your function report a failure by setting *ifail* to 1 and *fout* to *NaN*. The bound constraints and the possibility of function failure add some complexity to the poll step and the line search, and we will discuss that now.

Algorithm **poll_stencil_v1** is not aware of either bound or hidden constraints. Since the computation of the stencil derivative needs information of both failed points and points which do not satisfy the bound constraints, the poll step must report the successful points in the stencil. It is not difficult to extend **poll_stencil_v1** to handle these requirements. Algorithm **poll_stencil** is the generalization of Algorithm **poll_stencil_v1** that does those things. Note that the pruning of infeasible points happens inside the poll step, so that the function which computes the stencil gradient is only aware of the feasible directions, which are stored in the array V^{out}, and the function values at the feasible points in the stencil, which are stored in F^{out}. Algorithm **poll_stencil** also keeps track of the cost of the function evaluations. This algorithm differs from that in **imfil.m** only in that **imfil.m** can do function evaluations in parallel, and hence must check the bound constraints before doing a parallel evaluation to test the hidden constraints.

Algorithm 4.10.
$(f_{min}, x_{min}, F^{out}, V^{out}, icost) = $ **poll_stencil**(x, f, h, V)
   $icost = 0.$
   $j = 1.$
   for $i = 1, \ldots, K$ **do**
      if $L \leq x + hv_i \leq U$ **then**
         $[fout, ifail, icount] = f(x + hv_i).$
         if $ifail = 0$ **then**
            $F_j^{out} = fout, V_j^{out} = v_i, j = j + 1.$
         $icost = icost + icount.$

     end if
   end if
  end for
Find i^* so that $f(x + hv_{i^*}) = \min_i(F_i^{out})$.
$x_{min} = x + hv_{i^*}$, $f_{min} = f(x_{min})$.

After calling **poll_stencil**, one computes the stencil derivative by eliminating the infeasible points and using $\nabla f(x, V^{out}, h)$ in the inner iteration.

We must also modify the line search. For bound constraints we replace (4.10) with

$$f(\mathcal{P}(x - \lambda d)) < f(x). \tag{4.17}$$

The line search must be prepared for the evaluation to fail, in which case $f_{newt} =$ NaN and $ifail = 1$.

Algorithm 4.11.
$(f_{newt}, x_{newt}, acount) = $ **Armijo_linesearch**$(f, x, d, L, U, maxitarm)$
  $f_{base} = f(x)$.
  $x_{newt} = \mathcal{P}(x + d)$; $[f_{newt}, ifail, icost] = f(x_{newt})$.
  $acount = icost$.
  while $(f_{base} \leq f_{newt}$ or $ifail = 1)$ and $acount <= maxitarm$ **do**
    $d = \beta d$.
    $x_{newt} = \mathcal{P}(x + d)$; $[f_{newt}, ifail, icost] = f(x_{newt})$.
    $acount \leftarrow acount + icost$.
  end while

As we will see in the next section, we are at risk for terminating the optimization prematurely if we use a fixed stencil V and cannot evaluate f at all points in the stencil. **imfil.m** will let you add directions to the stencil and solve this problem.

4.4.4 Explicit Constraints

By an explicit constraint we mean one that can be evaluated by simply testing the variables and not by calling the function. Bound constraints are explicit constraints, but **imfil.m** handles them directly. We will express a system of P explicit constraints in the form

$$c(x) \leq 0, \tag{4.18}$$

where $c : R^N \rightarrow R^P$ and the inequality is understood componentwise.

One way to deal with explicit constraints is what [8] calls the **extreme barrier** approach. Here one simply flags a point as a failure if it is infeasible, i.e.,

$$(c(x))_i > 0 \text{ for any } 1 \leq i \leq P.$$

In the context of **imfil.m** one way to do this is to simply include a test for (4.18) into the code for f. If (4.18) fails, then set

$$[fout, ifail, icount] = [NaN, 1, 0].$$

The value $icount = 0$ indicates that there is no cost in testing an explicit constraint. You may think of this as making explicit constraints the same as zero-cost hidden constraints.

If one uses a fixed stencil and the extreme barrier approach for explicit constraints, then the optimization can stagnate at a nonoptimal point. Here is an example [5] of the problem. Suppose you seek to minimize

$$f(x) = 1 - (x)_2$$

on the set

$$\Omega = \{x \in R^2 \,|\, 0 \le (x)_i \le 1, 0 \le i \le 2\}$$

with the explicit constraint

$$(x)_1 + (x)_2 \le 1.$$

Suppose you start the optimization at

$$x_0 = (1/2, 1/2)^T$$

and use the stencil of positive and negative coordinate directions. For any $h < 1/2$ the stencil has two feasible points,

$$x_{left} = (1/2 - h, 1/2)^T \quad \text{and} \quad x_{down} = (1/2, 1/2 - h)^T,$$

neither of which will decrease f. The other two points on the stencil,

$$x_{right} = (1/2 + h, 1/2)^T \quad \text{and} \quad x_{up} = (1/2, 1/2 + h)^T,$$

satisfy the bound constraints but not the explicit constraint. The extreme barrier approach will miss the fact that

$$f(x_{up}) < f(x_0),$$

and one will get stencil failure for any $h < 1/2$. Therefore the iteration will stagnate. The solution is to add an extra direction v_{new} so that if $x_{new} = x_0 + h v_{new}$, then $f(x_{new}) < f(x_0)$. One can do this in **imfil.m** with the vstencil (see § 6.9.1) option as follows:

```
VS=[0 1; 0 -1; 1 0; -1 0; -1, .5]';
options=imfil_optset('vstencil',VS);
```

The code for this example lc_imfil.m in the Examples/Linear_Constraints directory.

We illustrate this example in Figure 4.2, where the entire image is the feasible set for the bound constraints and the shaded area in the upper right is the infeasible region for the explicit constraints. Here $v_{new} = (-1, 1/2)^T$, but any direction $v = (v_x, v_y)^T$ with $v_x < 0$ and $0 < v_y \le -v_x$ will work. We created Figure 4.2 with the MATLAB code linear_constraints.m in the Examples/Linear_Constraints directory.

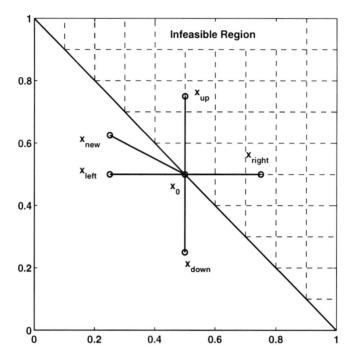

Figure 4.2. *Explicit constraints.*

In § 5.6 we will show that if one changes the stencil with each iteration in the proper way, then the direction set will be rich enough (at least asymptotically) to find a descent direction even with explicit or hidden constraints. In this particular example, the explicit constraints are linear, and adding the positive and negative tangent directions (i.e., $(1, -1)$ and $(-1, 1)$) [92] is an approach which, when managed carefully, generalizes to systems of linear constraints.

We describe the `add_new_directions` option in § 7.1. This option lets you add new directions via a function which you define. Your function uses the current point, current stencil, and the history of the iteration to enrich the stencil. You may also add $k > 0$ random directions to the stencil by setting the `random_directions` (see § 6.9.2) option to k,

```
options=imfil_optset('random_directions',k);
```

The `explore_function` is another way to enrich the search. See § 7.4 and § 7.4.1 for the details.

Chapter 5

Convergence Theory

To the practitioner, this chapter on convergence theory may be of less interest than the later chapters on examples. However, understanding some theory will help you solve algorithmic problems and better set the options.

We caution the reader that the theory for implicit filtering, and many other related methods [7, 38, 69, 75, 88, 105], relies in some way on assumptions of smoothness, near smoothness, or generalized smoothness. Convergence results are asymptotic, whereas methods like implicit filtering are generally neither run long enough for an asymptotic theory to be relevant nor applied to smooth problems. In practice, on the other hand, while the problems one actually wants to solve (and often does solve) may well not strictly satisfy the assumptions for the theory, the theory does provide useful guidance for algorithmic design, implementation, and use of the software.

In our theoretical results we will assume that the objective function f is a perturbation of a smooth function f_s,

$$f = f_s + \phi, \tag{5.1}$$

or that, for nonlinear least squares,

$$F = F_s + \phi. \tag{5.2}$$

We will call ϕ the **noise**. Implicit filtering and many other sampling methods can be regarded as enhanced versions of coordinate search, and the analysis for coordinate search is at the core of their convergence proofs.

In this chapter we will assume that $L_i = 0$ and $U_i = 1$ for all $1 \leq i \leq N$. This is consistent with the internal scaling that **imfil.m** does for you and makes the notation simpler.

5.1 Consequences of Stencil Failure

The two theorems in this section say that stencil failure is a good indicator that the first-order necessary conditions hold for f_s if the noise is sufficiently small. In § 5.2

we apply these results to prove two simple asymptotic convergence results.

We begin with the simple case in which S is the stencil of the positive and negative coordinate directions (4.1) and the problem is unconstrained. We define the local norm of the noise as

$$\|\phi\|_{S(x,h)} = \max_{z \in \{x\} \cup S(x,h)} |\phi(z)|.$$

Theorem 5.1. *Let (5.1) hold and let f_s be Lipschitz continuously differentiable. Assume that x is a point of stencil failure,*

$$f(x) = \min_{z \in S(x,h)} f(z).$$

Then

$$\|\nabla f_s(x)\| = O(h + \|\phi\|_{S(x,h)}/h). \qquad (5.3)$$

Proof. Stencil failure implies that

$$f_s(x) \le f_s(x \pm he_i) + 2\|\phi\|_{S(x,h)}.$$

Let L be the Lipschitz constant of ∇f. Then (see § 3.7)

$$-Lh/2 - 2\frac{\|\phi\|_{S(x,h)}}{h} \le \frac{f_s(x) - f_s(x - he_i)}{h} - \frac{Lh}{2} \le \frac{\partial f_s(x)}{\partial x_i}$$
$$\le \frac{f_s(x + he_i) - f_s(x)}{h} + \frac{Lh}{2} \le \frac{Lh}{2} + 2\frac{\|\phi\|_{S(x,h)}}{h}. \qquad (5.4)$$

Hence (5.3) holds. □

Corollary 5.2 follows from (5.4).

Corollary 5.2. *Let V be a set of directions which includes the positive and negative coordinate directions. Let (5.1) hold and let f_s be Lipschitz continuously differentiable. Then*

$$\nabla f(x, V, h) = \nabla f_s(x) + O(h + \|\phi\|_{S(x,h)}/h). \qquad (5.5)$$

Recall that in the bound constrained case

$$S(x, h) = \{z \mid z = x \pm he_i\} \cap \Omega.$$

We must assume that (after scaling) $h \le 1/2$, so that

$$x \pm he_i \in \Omega \text{ for at least one } i.$$

Theorem 5.3. *Let (5.1) hold and let f_s be Lipschitz continuously differentiable. Assume that x is a point of stencil failure,*

$$f(x) = \min_{z \in S(x,h)} f(z).$$

Then

$$x - \mathcal{P}(x - \nabla f_s(x)) = O(h + \|\phi\|_{S(x,h)}/h), \qquad (5.6)$$

Proof. In the bound constrained case we may only be able to compute a one-sided difference, and so the best we can do is to get a first-order accurate approximation of the sign of that derivative. In that event (5.3) will become

$$\pm \nabla f_s(x)^T e_i \leq O(h + \|\phi\|_{S(x,h)}/h) \qquad (5.7)$$

whenever

$$x \pm h e_i \in \Omega.$$

Assume, say, that $x + h e_i \in \Omega$ and $x - h e_i \notin \Omega$; hence

$$0 \leq (x)_i < 0 + h.$$

We apply (5.4) to obtain, for some $C > 0$,

$$\partial f_s(x)/\partial x_i \leq C(h + \|\phi\|_{S(x,h)}/h),$$

which implies that

$$\max(0, (x)_i - C(h + \|\phi\|_{S(x,h)}/h)) \leq (x - \nabla f_s(x))_i$$

$$\leq (x)_i + C(h + \|\phi\|_{S(x,h)}/h).$$

Hence,

$$(x)_i - \min((x)_i - 0, C(h + \|\phi\|_{S(x,h)}/h)) \leq (\mathcal{P}(x - \nabla f_s(x)))_i$$
$$\leq (x)_i + C(h + \|\phi\|_{S(x,h)}/h), \qquad (5.8)$$

which is equivalent to

$$(\mathcal{P}(x - \nabla f_s(x)))_i = (x)_i + O(h + \|\phi\|_{S(x,h)}/h). \qquad (5.9)$$

Since either (5.7) or (5.9) will hold for every i, the proof is complete. $\square$

As in the unconstrained case, the proof of Theorem 5.3 leads to an estimate for the accuracy of the projected stencil gradient.

Corollary 5.4. *Let V be a set of directions which includes the positive and negative coordinate directions. Let (5.1) hold and let f_s be Lipschitz continuously differentiable. Then*

$$\mathcal{P}(x - \nabla f(x, V, h)) = \mathcal{P}(x - \nabla f_s(x)) + O(h + \|\phi\|_{S(x,h)}/h). \qquad (5.10)$$

5.2   Stencil Failure and Coordinate Search

We will now prove a simple convergence result for the smooth f ($f = f_s$) and the unconstrained case.

Theorem 5.5. *Let f be Lipschitz continuously differentiable. Assume that the set $\{x \mid f(x) \leq f(x_0)\}$ is bounded. Let $\{x_n\}$ be the sequence of coordinate search iterations for the unconstrained problem*

$$\min f.$$

Then

$$\liminf_{n \to \infty} \|\nabla f(x_n)\| = 0.$$

Proof. Let h_n be the scale at the nth iteration. We first observe that $h_n \to 0$. This follows from the fact that for any fixed h

$$\{z \mid z = x \pm khe_i, k = 1, 2, \dots\}$$

has only finitely many points in the set $\{x \mid f(x) \leq f(x_0)\}$. Hence coordinate search will completely explore all the possibilities for a give scale in finitely many iterations, and must reduce the scale after that. Now let $\{x_{n_k}\}$ be the subsequence of $\{x_n\}$ of points of stencil failure. Then, by (5.3),

$$\lim_{k \to \infty} \nabla f(x_{n_k}) = O(h_{n_k}) \to 0,$$

which completes the proof. $\square$

For the bound constrained or noisy case, we obtain a similar result if we make the additional assumption that the noise decays as we approach optimality. We state and prove the result in the noisy bound constrained case.

Theorem 5.6. *Let V be a set of directions which includes the positive and negative coordinate directions. Let f satisfy (5.1) with f_s Lipschitz continuously differentiable in*

$$\Omega = \{x \in R^N \mid 0 \leq (x)_i \leq 1\}.$$

Let $\{x_n\}$ be the sequence of coordinate search iterations for the bound constrained problem. Assume that

$$\lim_{n \to \infty} \frac{\|\phi\|_{S(x_n, h_n)}}{h_n} = 0; \tag{5.11}$$

then

$$\liminf_{n \to \infty} \|x_n - \mathcal{P}(x_n - \nabla f_s(x_n))\| = 0. \tag{5.12}$$

Proof. As in the proof of Theorem 5.5 we let $\{x_{n_k}\}$ be the subsequence of $\{x_n\}$ of points of stencil failure. Equations (5.11) and (5.6) imply

$$\lim (x_{n_k} - \mathcal{P}(x_{n_k} - \nabla f_s(x_{n_k}))) = 0,$$

which completes the proof. $\square$

5.3   Positive Spanning Sets and Positive Bases

One might ask if the convergence theory is valid if there are fewer than $2N$ points in the stencil, and how many points one must use. In the unconstrained case, there is a precise answer. However, this answer does not apply to the bound constrained case and becomes another example of the need to enrich the stencil.

The proof of the coordinate search results in § 5.2 requires that, at a minimum, V be a **positive spanning set** [38, 88, 91, 134]. By this we mean that any vector in $w \in R^N$ can be written (possibly not in a unique way) as

$$w = \sum_{i=1}^{M} a_i v_i,$$

where $a_i \geq 0$. In this section, we will assume, for convenience, that the columns of V have norm 1.

We will show that stencil failure with a positive spanning set of directions is sufficient to conclude that $\|\nabla f\| = O(h)$. To quantify this we define the **condition number** of the positive spanning set by

$$\kappa(V) = \sqrt{N} \min \|A\|_\infty$$

subject to the constraints that the $2N \times M$ matrix A have nonnegative entries and that $AV^T = [U, -U]^T$ for some orthogonal matrix U.

Theorem 5.7. *Let f satisfy (5.1). Let ∇f_s be Lipschitz continuous with Lipschitz constant L. Let V be a positive spanning set. Then stencil failure implies that*

$$\|\nabla f_s(x)\| \leq \kappa(V) \left(\frac{Lh}{2} + \frac{\|\phi\|_{S(x,h)}}{h} \right). \tag{5.13}$$

Proof. Let

$$\mathbf{1} = (1, 1, \ldots, 1)^T.$$

Stencil failure implies (see § 3.7) that

$$V^T \nabla f_s(x) \leq \frac{Lh}{2} + \frac{\|\phi\|_{S(x,h)}}{h} \mathbf{1}, \tag{5.14}$$

where the inequality in (5.14) is understood componentwise.

Now, let U be an $N \times N$ orthogonal matrix such that

$$AV^T = W \equiv [U, -U]$$

for which the matrix A with nonnegative entries satisfies $AV^T = W$ and $\kappa(A) = \sqrt{N}\|A\|_\infty$. Since A has nonnegative entries, we see that (5.14) implies that

$$W^T \nabla f_s(x) = AV^T \nabla f_s(x) \leq A\mathbf{1} \left(\frac{Lh}{2} + \frac{\|\phi\|_{S(x,h)}}{h} \right) \leq \|A\|_\infty \left(\frac{Lh}{2} + \frac{\|\phi\|_{S(x,h)}}{h} \right) \mathbf{1}. \tag{5.15}$$

Since the columns of $W^T \nabla f$ include both $\partial f / \partial u_j$ and $-\partial f / \partial u_j$ for $1 \leq j \leq N$, we have

$$\|\nabla f_s(x)\| \leq \sqrt{N} \|A\|_\infty \left(\frac{Lh}{2} + \frac{\|\phi\|_{S(x,h)}}{h} \right). \tag{5.16}$$

Note that if $V = [I, -I]$ is the default stencil, we may use $U = V$ and $\kappa(V) = 1$. □

A direct corollary of the proof of Theorem 5.7 is an estimate for $DF(x, V, h) - F_s'(x)$.

Corollary 5.8. *Let* (5.2) *hold. Assume that* $F_s : R^N \to R^M$ *is Lipschitz continuously differentiable. Let* V *be a positive spanning set. Then*

$$\|F_s'(x) - DF(x, V, h)\| \leq \kappa(V) \left(\frac{\gamma h}{2} + \frac{\|\phi\|_{S(x,h)}}{h} \right), \tag{5.17}$$

where γ *is the Lipschitz constant of* F'.

A **positive basis** is a positive spanning set which has no positive spanning subsets. No positive basis can have fewer than $N + 1$ directions, hence the theory requires at least $N + 1$ points in the stencil. An example of such a set is $V_P = [I, -1/\sqrt{N}]$. The default stencils in **imfil.m** are $[I, -I]$, which is a positive basis stencil with $2N$ elements, I, and V_P, with the first of these being the default and the recommended choice of the three.

The stencil must contain $[I, -I]$ to make the convergence theory work with bound constraints. The positive basis stencil V_P can easily miss descent directions. To see this define

$$f(x) = x_1 + 10 * (1 - x_2)$$

on the set

$$\Omega = \{x \in R^2 \,|\, 0 \leq (x)_i \leq 1, 0 \leq i \leq 2\}.$$

If you start the search at $x = (1, 1)^T$, the only feasible direction from the positive basis stencil will be $v = -1/\sqrt{2}$, which is a direction of increase for f since

$$\partial f(x) / \partial v = 9.$$

We encountered a similar problem in § 4.4.4 when a search with the default stencil also stagnated at a nonoptimal point. To avoid this, our results for bound-constrained problems will require that V be a positive spanning set which includes the positive and negative coordinate directions. In the presence of either hidden or explicit constraints, we must also enrich the stencil with enough directions to avoid the kinds of stagnation we have seen in this section and in § 4.4.4. As we do that in § 5.6, we will find that we can also relax our smoothness requirements on f_s.

5.4 A Convergence Theorem for Implicit Filtering

We are now ready to state and prove two basic convergence theorems for implicit filtering. Pure sampling algorithms like Hooke–Jeeves [68], coordinate search, and the many variations of multidirectional search [7, 45, 88] search on a pattern which fills space. Our analysis of coordinate search exploits this property with the observation that we can sample only finitely many points at a given scale h before we must shrink the stencil. The implicit filtering iteration does not remain on a grid, and hence we must make assumptions on the way in which the inner iteration Algorithm **imfil_inner** terminates.

We will state results for unconstrained optimization, where any positive spanning set will do for the stencil, and for bound constrained optimization, where one must use the positive and negative coordinate directions to avoid the problem described at the end of § 5.3, where the direction set is not rich enough to capture a descent direction. The results for nonlinear least squares are similar.

When we discuss implicit filtering in the unconstrained case, we mean Algorithm **imfil_outer** with $U = \infty$ and $L = -\infty$.

Theorem 5.9. *Let f satisfy (5.1). Assume that ∇f_s is Lipschitz continuous. Let V be a positive spanning set and let $\{x_n\}$ be the implicit filtering iterations for the unconstrained problem*

$$\min f.$$

Assume that (5.11) holds and that the inner iteration terminates infinitely often with either stencil failure or when $\|\nabla f(x, V, h)\| < \tau h$. Then

$$\liminf_{n \to \infty} \|\nabla f_s(x_n)\| = 0. \tag{5.18}$$

Proof. The proof is a direct application of Theorem 5.7 and Corollary 5.8. Let $\{x_{n_j}\}$ be an infinite subsequence of $\{x_n\}$ for which the inner iteration terminates either with stencil failure or when $\|\nabla f(x, V, h)\| < \tau h$. Then one may apply either the theorem or corollary to conclude that

$$\|\nabla f_s(x_{n_j})\| = O\left(h_{n_j} + \frac{\|\phi\|_{S(x, h_{n_j})}}{h_{n_j}}\right).$$

Since $h_{n_j} \to 0$, (5.11) implies (5.18). □

Theorem 5.10. *Let f satisfy (5.1) with f_s Lipschitz continuously differentiable in*

$$\Omega = \{x \in R^N \,|\, 0 \le (x)_i \le 1\}.$$

Let V be a positive spanning set which includes the positive and negative coordinate directions. Let $\{x_n\}$ be the sequence of implicit filtering iterates. Assume that (5.11) holds and that the inner iteration terminates infinitely often with either stencil failure or when $\|x - \mathcal{P}(x - \nabla f(x, V, h))\| < \tau h$. Then

$$\liminf_{n \to \infty} \|x_n - \mathcal{P}(x_n - \nabla f_s(x_n))\| = 0. \tag{5.19}$$

Proof. The proof follows from Theorem 5.3 and Corollary 5.4. Since at any of the infinitely many points $\{x_{n_j}\}$ at which there is either stencil failure or at which $\|x_{n_j} - \mathcal{P}(x_{n_j} - \nabla f(x_{n_j}, V, h_{n_j}))\| < \tau h_{n_j}$, h will be reduced and

$$\|x_{n_j} - \mathcal{P}(x_{n_j} - \nabla f_s(x_{n_j}))\| = O(h_{n_j} + \|\phi\|_{S(x, h_{n_j})}/h_{n_j}).$$

This completes the proof.   □

5.5   Convergence Rates

When one analyzes the convergence rates of an iterative method for bound constrained optimization, the usual procedure is first to show that the binding constraints are identified after finitely many steps, and then to invoke the theory for unconstrained problems. We will do this, under more restrictive assumptions on h and ϕ than we have used before, in § 5.5.1. Once the binding constraints have been identified, the convergence rate study reverts to the local convergence theory of the underlying method for unconstrained problems [12, 75].

We will now describe a few local convergence rate results for unconstrained problems. Our convergence results so far have assumed that (5.1) holds, that f_s is Lipschitz continuously differentiable, and that (5.11) holds. Convergence rates require a stronger form of (5.11) and a rapidly decreasing sequence of scales. In § 5.5.2 we will illustrate this with a proof of an r-linear convergence result. This result has weaker assumptions than those from [76]. For results for the optimization case and superlinear convergence, see [30, 76].

In all these results one must connect the noise ϕ and the scale h to the error in the solution. One can actually do this in some applications, as we illustrate in Chapter 8.

5.5.1   Identification of the Binding Constraints

Let $\{x_n\}$ denote the implicit filtering iterations. In this section we will assume that the stencil V contains the coordinate directions and that (5.1) holds.

We seek conditions on ϕ and the sequences $\{h_n\}$ and $\{\epsilon_n\}$ under which convergence of $\{x_n\}$ to x^*, a minimizer of f_s which satisfies the second-order sufficiency conditions, implies that

$$\overline{\mathcal{B}}^{\epsilon_n}(x_n, V, h_n) = \{i \,|\, U_i - (x)_i \le \epsilon \text{ and } \nabla f(x, V, h)^T e_i \le -\sqrt{\epsilon}\}$$

$$\cup \{i \,|\, (x)_i - L_i \le \epsilon \text{ and } \nabla f(x, V, h)^T e_i \ge \sqrt{\epsilon}\} \qquad (5.20)$$

$$= \mathcal{B}_s(x^*)$$

for n sufficiently large.

In (5.20), $\mathcal{B}_s(x)$ is the ϵ-binding set at x relative to f_s and $\mathcal{B}_s(x^*)$ is the binding set at x^* relative to f_s. Using the notation of § 3.5.2,

$$\mathcal{B}_s^\epsilon(x) = \mathcal{B}_s^\epsilon(x, 1),$$

where, for $\alpha > 0$,

$$\mathcal{B}_s^\epsilon(x, \alpha) = \{i \,|\, 1 - (x)_i \le \epsilon \text{ and } \partial f_s(x)/\partial(x)_i \le -\alpha\sqrt{\epsilon}\}$$

$$\cup \{i \,|\, (x)_i - 0 \le \epsilon \text{ and } \partial f_s(x)/\partial(x)_i \ge \alpha\sqrt{\epsilon}\}.$$

We repeat the remark from § 4.3.2 that the theory is expressed in terms of $\overline{\mathcal{B}}^\epsilon(x, V, h)$ rather than

$$\mathcal{B}^\epsilon(x, V, h) = \{i \mid U_i - (x)_i \le \epsilon\} \cup \{i \mid (x)_i - L_i \le \epsilon\},$$

whereas the implementation uses $\mathcal{B}^\epsilon(x, V, h)$. Of course, if all the active constraints at x^* are binding, then the theory may use $\mathcal{B}^\epsilon(x, V, h)$ as well.

Theorem 5.11 is our result on identification of $\mathcal{B}$ for optimization problems. The result for nonlinear least squares problems is similar, with the only change being the error estimate in the difference gradient (5.22).

Theorem 5.11. *Assume that the stencil V contains the coordinate directions and that (5.1) holds. Assume that ∇f_s is Lipschitz continuous with Lipschitz constant L. Let $\{x_n\}$ be the implicit filtering iterates. Assume that $x_n \to x^*$, where x^* is a local minimizer of f_s in Ω which satisfies the second-order sufficiency conditions. Assume that (5.11) holds, that $h_n \to 0$, and that there is $\alpha \in (0,1)$ such that*

$$\tau_n \equiv \frac{Lh_n}{2} + 2\frac{\|\phi\|_{S(x,h_n)}}{h_n} \le (1-\alpha)\sqrt{\epsilon_n} \tag{5.21}$$

for all n. Then there is $\bar{\epsilon} > 0$ so that if $\epsilon_n \le \bar{\epsilon}$ for all n sufficiently large,

$$\overline{\mathcal{B}}^{\epsilon_n}(x_n, V, h_n) = \mathcal{B}_s(x^*)$$

for all n sufficiently large.

Proof. The proof of Theorem 5.1 and Corollary 5.4 imply that

$$\|\nabla f(x_n, V, h_n)) - \nabla f_s(x_n)\| \le \tau_n, \tag{5.22}$$

where τ_n is defined by (5.21).

Our assumptions and Lemma 3.13 imply that if $\bar{\epsilon}$ is sufficiently small, then

$$\mathcal{B}_s^{\epsilon_n}(x^*) = \mathcal{B}_s^{\epsilon_n}(x^*, \alpha) = \mathcal{B}_s^{\bar{\epsilon}}(x^*) = \mathcal{B}_s^{\bar{\epsilon}}(x^*, \alpha) = \mathcal{B}_s(x^*).$$

Let n be large enough so that $\epsilon_n \le \bar{\epsilon}$. Assume that $i \in \overline{\mathcal{B}}^{\epsilon_n}(x_n, V, h_n)$. Then

$$\nabla f(x_n, V, h_n)^T e_i < -\sqrt{\epsilon_n},$$

which by (5.22) implies that

$$\nabla f_s(x_n) \le \tau_n - \sqrt{\epsilon_n} \le -\alpha\sqrt{\epsilon_n}.$$

So $i \in \mathcal{B}_s^{\epsilon_n}(x^*, \alpha) = \mathcal{B}_s(x^*)$. Hence

$$\overline{\mathcal{B}}^{\epsilon_n}(x_n, V, h_n) \subset \mathcal{B}_s(x^*).$$

Conversely, let $i \in \mathcal{B}_s(x^*)$. Without loss of generality we may assume that $1 = (x^*)_i$ and $\partial f_s(x^*)/\partial(x)_i = -d_i < 0$. Let n be large enough so that

$$\nabla f_s(x_n) < -d_i/2 < -2\sqrt{\epsilon_n}.$$

Then
$$\nabla f(x_n, V, h_n)^T e_i < -d_i + \tau_n < -d_i/2 + (1-\alpha)\sqrt{\epsilon_n} < -\sqrt{\bar{\epsilon}}$$
for n sufficiently large. Hence
$$i \in \overline{\mathcal{B}}^{\bar{\epsilon}}(x_n, V, h_n).$$

This completes the proof. $\square$

In **imfil.m** we set $\epsilon_n = \bar{\epsilon} \equiv 10^{-6}$. The argument for this is that Theorem 5.11 asserts that the binding constraints are identified if $\bar{\epsilon}$ is sufficiently small, so we may reasonably assume, given the internal scaling in **imfil.m**, that 10^{-6} is "sufficiently small."

Another approach, which has more in common with the algorithm for smooth problems, would be to set
$$\epsilon_n = \|x_n - \mathcal{P}(x_n - \nabla f(x_n, V, h_n))\|. \tag{5.23}$$

In order to apply Theorem 5.11 one would have to manage $\{h_n\}$ and $\{\epsilon_n\}$ so that (5.21) holds. Adding (5.23) to (5.21), we obtain
$$\epsilon_n = \|x_n - \mathcal{P}(x_n - \nabla f(x_n, V, h_n))\| \le \tau_n = \frac{Lh_n}{2} + 2\frac{\|\phi\|_{S(x,h_n)}}{h_n}$$
$$\le (1-\alpha)\sqrt{\epsilon_n}.$$

In practice we find that $x_n - \mathcal{P}(x_n - \nabla f(x_n, V, h_n))$ converges to zero slowly, so (5.23) does not have the desirable properties it does in the smooth case.

5.5.2 Gauss–Newton Iteration

The Gauss–Newton has enough special structure to allow one to derive a strong local convergence theory [76]. We will let $F : R^N \to R^M$ with $M \ge N$ and assume that (5.2) holds. We compare the implicit filtering Gauss–Newton iteration to the Gauss–Newton iteration for the smooth problem
$$\min \|F_s(x)\|^2/2 \tag{5.24}$$
and will assume that Assumption 1 holds for F_s. We will need a trivial consequence of Assumption 1 for our analysis, which we state as a lemma.

Lemma 5.12. *Let $F : R^N \to R^M$ with $M \ge N$ and assume that (5.2) holds. Let Assumption 1 hold for F_s. Then there are $\bar{\sigma}_1 \ge \bar{\sigma}_N > 0$ and $\rho > 0$ so that if*
$$x \in \mathcal{N}(\rho) \equiv \{x \mid \|x - x^*\| < \rho\}, \tag{5.25}$$
the singular values of $F_s'(x)$ are in the interval $[\bar{\sigma}_N, \bar{\sigma}_1]$.

We will assume that $\kappa(V) = 1$, so we may write (5.17) as
$$DF(x, V, h) = F_s'(x) + E_J(x), \tag{5.26}$$

where

$$\|E_J(x)\| \leq \left(\frac{\gamma h}{2} + \frac{\|\phi\|_{S(x,h)}}{h}\right). \tag{5.27}$$

In (5.27) γ is the Lipschitz constant of F_s'. We can use (5.26) and (5.27) to estimate the error in the stencil gradient

$$E_g(x) \equiv DF(x, V, h)^T F(x) - F_s'(x)^T F_s(x) \tag{5.28}$$

by

$$\|E_g(x)\| \leq \|F_s'(x)\|\|\phi\|_{S(x,h)} + \|E_J(x)\|\|F(x)\|. \tag{5.29}$$

Let $\rho > 0$ and let

$$C_G = \max_{x \in \mathcal{N}(\rho)} (\|F_s'(x)\| + \|F(x)\|).$$

Then (5.27) and (5.29) imply that

$$\|E_g(x)\| \leq C_G \left(\frac{\gamma h}{2} + \frac{\|\phi\|_{S(x,h)}}{h}\right) \tag{5.30}$$

for all $x \in \mathcal{N}(\rho)$.

Now let $x \in \mathcal{N}(\rho)$. We wish to compare the implicit filtering Gauss–Newton step

$$s = -(DF(x, V, h)^T DF(x, V, h))^{-1} DF(x, V, h)^T F(x)$$

with the Gauss–Newton step for F_s,

$$s^{GN} = -(F_s'(x)^T F_s'(x))^{-1} F_s'(x)^T F_s(x),$$

and to determine conditions on h and ϕ for which the two steps are close enough to enable us to develop a local convergence theory.

As a first step we give conditions for nonsingularity of $DF(x, V, h)^T DF(x, V, h)$.

Lemma 5.13. *Let $F : R^N \to R^M$ with $M \geq N$ and assume that (5.2) holds. Let Assumption 1 hold for F_s. Assume that $x \in \mathcal{N}(\rho)$ and that*

$$E_J^{max}(h) \equiv \frac{\gamma h}{2} + \frac{\|\phi\|_{S(x,h)}}{h} < \sqrt{\bar{\sigma}_1^2 + \bar{\sigma}_N^2/4} - \bar{\sigma}_1. \tag{5.31}$$

Then $DF(x, V, h)^T DF(x, V, h)$ is nonsingular and

$$(DF(x, V, h)^T DF(x, V, h))^{-1} = (F_s'(x)^T F_s'(x))^{-1} + E_N, \tag{5.32}$$

where

$$\|E_N\| \leq \min\left(\frac{3}{16\bar{\sigma}_N^2}, \frac{9E_J^{max}(h)\bar{\sigma}_1}{4\bar{\sigma}_N^4}\right).$$

Proof. Since $x \in \mathcal{N}(\rho)$, $F_s'(x)$ has full column rank. So, using (5.26) and Lemma 5.12, we have

$$\|(F_s'(x)^T F_s'(x))^{-1}(DF(x, V, h)^T DF(x, V, h)) - I\| \leq \delta_J, \tag{5.33}$$

where
$$\delta_J = \bar{\sigma}_N^{-2} E_J^{max}(h)(2\bar{\sigma}_1 + E_J^{max}(h)) < 1/4 \qquad (5.34)$$

by (5.27) and (5.31). Therefore, $(F_s'(x)^T F_s'(x))^{-1}$ is an approximate inverse of $DF(x,V,h)^T DF(x,V,h)$. This implies that $DF(x,V,h)^T DF(x,V,h)$ is nonsingular by the Banach lemma [74]. Hence, (5.32) holds with

$$\|E_N(x)\| \le \frac{\|(F_s'(x)^T F_s'(x))^{-1}\|\delta_J}{1 - \delta_J} \le \frac{3\bar{\sigma}_N^{-2}}{16}. \qquad (5.35)$$

Now, (5.31) implies that $\|E_J\| < \bar{\sigma}_N/2 < \bar{\sigma}_1$, so

$$\delta_J \le \frac{3\bar{\sigma}_1\|E_J\|}{\bar{\sigma}_N^2},$$

which combined with (5.35) completes the proof. □

We now combine Lemmas 5.12 and 5.13 to obtain an estimate for the difference between s^{GN}, the Gauss–Newton step for F_s, and the implicit filtering Gauss–Newton step s.

Corollary 5.14. *Let $x \in \mathcal{N}(\rho)$ and let the assumptions of Lemma 5.13 hold. Then*

$$\|s^{GN} - s\| \le C_S E_J^{max}(h). \qquad (5.36)$$

Recall that if the residual $F_x(x^*)$ is sufficiently small, the Gauss–Newton iteration will converge q-linearly and (3.66) will hold for $x_+ = x_c + s^{GN}$, i.e.,

$$\|x^{GN} - x^*\| \le \sigma\|x - x^*\|$$

for some $\sigma \in (0,1)$.

We can now prove convergence and establish an r-linear convergence rate for the implicit filtering Gauss–Newton iteration. To do this we must connect the size of the noise, the scale, and the error $x - x^*$.

Theorem 5.15. *Let the assumptions of Lemmas 5.12 and 5.13 hold. Assume that x^* is a limit point of the implicit filtering sequence $\{x_n\}$ and that*

$$E_J^{max}(h_n) = h_n + \frac{\|\phi\|_{S(x,h_n)}}{h_n} \le C_\Phi \bar{r}^n \qquad (5.37)$$

for some $r \in (0,1)$. Assume that the Gauss–Newton iteration for F_s is locally q-linearly convergent to x^* with q-factor σ. Then $x_n \to x^*$ and there is $C_R > 0$ such that

$$\|x_n - x^*\| \le C_R \max\left(\bar{r}, \frac{1+\sigma}{2}\right)^n \qquad (5.38)$$

for all n sufficiently large.

Proof. Without loss of generality we may assume that

$$\bar{r} \ge \frac{1+\sigma}{2}.$$

Let ρ be small enough so that the conclusions of Lemma 5.12 hold and that

$$\|x^{GN} - x^*\| \le \sigma\|x - x^*\|$$

for all $x \in \mathcal{N}(\rho)$. Let n be such that $x_n \in \mathcal{N}(\rho)$ and

$$AC_\Phi \bar{r}^n \le \rho,$$

where $A = 2/(1 - \sigma)$.

Let

$$r = \max\left(\bar{r}, \frac{1 + \sigma}{2}\right).$$

We will show that

$$\|x_{n+1} - x^*\| \le r \max\left(\|x_n - x^*\|, AC_\Phi r^n\right), \tag{5.39}$$

which will imply that for all $k \ge 0$

$$\|x_{n+k} - x^*\| \le r^k \max\left(\|x_n - x^*\|, AC_\Phi r^n\right),$$

which, in turn, will complete the proof.

Corollary 5.14 and our assumptions imply that

$$\|x_{n+1} - x^*\| \le \|x_{n+1}^{GN} - x^*\| + C_\Phi r^n \le \sigma\|x_n - x^*\| + C_\Phi r^n. \tag{5.40}$$

We will consider two cases. If

$$\|x_n - x^*\| \le AC_\Phi r^n,$$

then (5.40) implies that

$$\|x_{n+1} - x^*\| \le (1 + A\sigma)C_\Phi r^n \le \frac{1 + A\sigma}{r} C_\Phi r^{n+1} \le AC_\Phi r^{n+1}$$

since $r \ge (1 + \sigma)/2$ and

$$\frac{1 + A\sigma}{r} \le \frac{2 + 2A\sigma}{(1 + \sigma)} = A.$$

So in this case (5.39) holds.

On the other hand, if $\|e_n\| \ge AC_\Phi r^n$, then (5.40) implies that

$$\|x_{n+1} - x^*\| \le \left(\sigma + \frac{1}{A}\right)\|x_n - x^*\|$$

$$= \frac{1 + \sigma}{2}\|x_n - x^*\| \le r\|x_n - x^*\|.$$

This completes the proof. $\square$

The least believable of the assumptions for Theorem 5.15 is (5.37). This is really an assumption that the noise in f can be managed in a quantifiable way. In some cases this assumption is realistic, however. In Chapter 8 we consider a problem in which the MATLAB initial value problem solver ode15s is embedded in f. ode15s has step size and error control, and hence one can attempt to change the error in f as a function of h using the options to ode15s and the scale_aware option. See Chapter 8 (§ 8.3) for guidance.

5.6   Enriched Stencils

In § 5.2 and § 5.4 we showed that if (5.11) holds and the noise decays sufficiently rapidly as the iteration progresses, then any limit point of the coordinate search or implicit filtering iteration satisfies the necessary conditions for optimality. We will generalize those results here to show how one can modify the stencils to obtain convergence even if there are hidden and/or explicit constraints. The stencils must change as the iteration progresses if we are to conclude that convergence of implicit filtering (or coordinate search, for that matter) implies that some form of the first-order necessary conditions holds. We will cast our discussion in the language of hidden constraints, keeping in mind that one source of hidden constraints could be an application of the extreme barrier method. The analysis in this section is taken from [49].

The problem we must address is that if f fails to return a value for some values of $x \in \Omega$, as it would if we had hidden constraints or were using the extreme barrier approach to handle explicit constraints, then (5.11) may not hold. In that case, as the example in § 4.4.4 illustrates, either implicit filtering or coordinate search can stagnate at a point which does not satisfy the necessary conditions.

A secondary purpose of this section is to draw the contrast between the Taylor expansions we use to analyze optimization algorithms for differentiable problems and the more general approximations which must be used for functions which are Lipschitz continuous but not differentiable.

Smooth Problems and Regular Feasible Sets

We will let $\mathcal{D} \subset \Omega$ be the feasible set for all hidden constraints, i.e., the subset of Ω for which f returns a value. In this section we will assume that f satisfies (5.1) and that f_s is Lipschitz continuously differentiable.

Throughout this chapter we will assume that $\mathcal{D}$ is regular in the sense that for any $x \in \mathcal{D}$ the **Clarke cone** or **tangent cone** to $\mathcal{D}$ at x [31],

$$T_{\mathcal{D}}^{Cl}(x) = \mathrm{cl}\{v \mid x + tv \in \mathcal{D} \text{ for all sufficiently small } t > 0\}, \qquad (5.41)$$

is the closure of its nonempty interior. Here $\mathrm{cl}(Z)$ denotes the closure of a set $Z \subset R^N$. We will explain the need for this technical assumption in more detail when we relax our smoothness assumptions on f_s. We do not make any further assumptions on $\mathcal{D}$ and only assume that points outside of $\mathcal{D}$ are flagged as failed points so that that stencil failure is declared if the current point is the best feasible point on the stencil.

We express the necessary conditions for optimality in the general form [31]

$$\partial f_s(x^*)/\partial v \geq 0 \text{ for all } v \in T_{\mathcal{D}}(x). \qquad (5.42)$$

If $\mathcal{D}$ is determined by smooth inequality constraints, then (5.42) is equivalent to the standard necessary conditions for inequality constrained problems [31, 51].

If $\mathcal{D} = \Omega$, then (5.42) is equivalent to (3.20) or (3.21). In the smooth case one can verify the necessary conditions by evaluating $\partial f_s/\partial v$ in the positive and negative coordinate directions. Recall that the proof of Theorem 5.10 depends on the inclusion of these directions in the stencil.

In the example in § 4.4.4 we do not have enough directions in the stencil to detect that the iteration has stagnated at a nonoptimal point. The reason for this is that the tangent cone at the stagnation point contains descent directions which cannot be detected by the simple stencil of positive and negative coordinate directions. In the example in § 4.4.4 the solution was to add directions that were close enough to the tangent cone and thereby find a descent direction. In fact, if $\mathcal{D}$ is determined by smooth explicit constraints, one could compute the tangent vectors (the constraint gradients) for the active constraints near the current iterate and add those to the direction set. This is the strategy adopted in [92] for linear constraints. The `add_new_directions` option will let you do this (see § 7.1).

In the general case, however, we do not have knowledge of the constraints and are at risk of the kind of failure we saw in § 4.4.4 if we use the same stencil for each iteration. The solution is to let the stencil vary as the iteration progresses in a way that, at least in the limit, will produce directions which get arbitrarily near all directions in $T_{\mathcal{D}}$.

To formalize this discussion, we must first define what we mean by a sufficiently rich sequence of directions. We use the notation from [49]. We assume that the direction set for the stencil in (4.4) changes with n, so the nth stencil is

$$S(x_n, h_n) = S(x_n, h_n, V_n) = \{z \,|\, z = x_n + h_n v_i, \, v_i \in V_n \, 1 \le i \le K_n\}, \qquad (5.43)$$

and the nth direction set is

$$V_n = \{v_i^n\}_{i=1}^{K_n}. \qquad (5.44)$$

We let $\mathcal{V} = \{V_n\}$ be the sequence of direction sets. We say that $\mathcal{V}$ is **rich** if for any unit vector $v \in R^N$ and any subsequence $\mathcal{W} = \{W_{n_j}\}$ of $\mathcal{V}$

$$\liminf_{j \to \infty} \min_{w \in W_{n_j}} \|w - v\| = 0. \qquad (5.45)$$

In **imfil.m** we generate rich direction sets by adding random directions to a base stencil. You set the `random_stencil` option to k to add k random directions to the stencil. If $k \ge 1$ this will, with probability one, produce a rich sequence of directions. You can also make rich sequences of directions deterministically. See [7] for a clever way to do this.

The analysis in this section applies to many methods, including implicit filtering, coordinate search, and other sampling methods [7]. We consider a sequence of iterations $\{x_n, V_n\}$, where x_n approximates a minimizer of f_s. We sample f on $S(x_n, h_n, V_n)$ and reduce the scale h_n if stencil failure occurs. We will also assume that a generalization of (5.11) holds:

$$\lim_{n \to \infty} \frac{\|\phi\|_{S(x_n, h_n, V_n)}}{h_n} = 0. \qquad (5.46)$$

Rich sequences of stencils are essentially, by definition, what is needed for stencil failure to be a signal of optimality. Theorem 5.16 makes this precise

Theorem 5.16. *Let $\mathcal{V}$ be a rich sequence of stencils. Let f satisfy (5.1), where f_s is Lipschitz continuously differentiable and ϕ satisfies (5.46). Let f be defined on $\mathcal{D} \subset \Omega$ where $T_{\mathcal{D}}(x)$ is the closure of its interior for all $x \in \mathcal{D}$. Let $\{x_n\} \subset \mathcal{D}$ be a sequence of iterations such that stencil failure occurs for all but finitely many n. Then any limit point of the sequence satisfies (5.42).*

Proof. Let x^* be a limit point of $\{x_n\}$ and let $v \in T_{\mathcal{D}}(x^*)$. Then, since $\mathcal{V}$ is rich, there is a subsequence, which we will denote by $\{x_{n_j}\}$ and $v_{n_j} \in V_{n_j}$ such that

- $x_{n_j} \to x^*$ and

- $v_{n_j} \to v$.

Stencil failure and (5.46) then imply that

$$\partial f_x(x^*)/\partial v \geq 0,$$

which completes the proof. □

Nonsmooth f_s

The analysis above can be extended beyond Theorem 5.16 in two directions. First we may relax the smoothness conditions on f_s to require only Lipschitz continuity. Second, we can extend the notion of tangent cone to require less regularity of $\mathcal{D}$ [6, 7, 31, 113]. Much of this section is based on [31], which is the primary reference for nonsmooth analysis. We will not present the results from [7, 49] in full generality and will continue to assume that $\mathcal{D}$ satisfies the assumptions of Theorem 5.16. We will present a simple result for Lipschitz continuous f_s to illustrate the ideas.

We will continue to assume that the tangent cone is the closure of its nonempty interior. This is a regularity assumption on the boundary of $\mathcal{D}$. We will also assume that $\mathcal{V}$ is rich and that f satisfies (5.1), but not that f_s is differentiable. Instead we will assume that f_s is Lipschitz continuous in $\mathcal{D}$. For Lipschitz continuous functions f_s we may define the **generalized directional derivative** of f_s at a point $x \in \mathcal{D}$ in the direction v as

$$f_s^{\circ}(x; v) = \limsup_{\substack{y \to x, \ y \in \mathcal{D} \\ t \downarrow 0, \ y+tv \in \mathcal{D}}} \frac{f_s(y + tv) - f_s(y)}{t}. \tag{5.47}$$

The generalization of the necessary conditions (5.42) is simply that

$$f_s^{\circ}(x^*; v) \geq 0 \text{ for all } v \in T_{\mathcal{D}}^{Cl}(x^*). \tag{5.48}$$

The convergence theorem [49] is very close to Theorem 5.16, but its proof needs a bit more care.

Theorem 5.17. *Let $\mathcal{V}$ be a rich sequence of stencils. Let f satisfy (5.1), where f_s is Lipschitz continuous and ϕ satisfies (5.46). Let f be defined on $\mathcal{D} \subset \Omega$ where $T_{\mathcal{D}}(x)$ is the closure of its interior for all $x \in \mathcal{D}$. Let $\{x_n\} \subset \mathcal{D}$ be a sequence of* *iterations such that stencil failure occurs for all but finitely many n. Then any limit point of the sequence satisfies* (5.48).

Proof. By taking subsequences if necessary, we may assume that $x_n \to x^*$ and there are directions $v_n \in V_n$ with $v_n \to v \in T_{\mathcal{D}}^{Cl}(x^*)$. The definition of the generalized directional derivative and the convergence of x_n to x^* imply that

$$f_s^o(x^*; v) \geq \lim_{n \to \infty} \frac{f_s(x_n + h_n v) - f_s(x_n)}{h_n}. \tag{5.49}$$

Let l_s denote the Lipschitz constant of f_s. Since

$$f_s(x_n + h_n v) \leq f_s(x_n + h_n v_n) + l_s h_n \|v - v_n\|,$$

we have

$$f_s(x_n + h_n v) - f_s(x_n) \geq f_s(x_n + h_n v_n) - f_l(\xi_n) - l_s h_n \|v_n - v\|$$
$$= f_s(x_n + h_n v_n) - f_s(x_n) + o(h_n). \tag{5.50}$$

In view of (5.46), (5.50) implies that

$$f_s(x_n + h_n v_n) - f_s(x_n) \geq f(x_n + h_n v_n) - f(x_n) - 2\|\phi\|_{S(x_n, h_n)}$$
$$= f(x_n + h_n v_n) - f(x_n) + o(h_n). \tag{5.51}$$

Since stencil failure occurs at all but finitely many x_n, we combine (5.50) and (5.51) to obtain

$$\frac{f_l(\xi_n + h_n v) - f_l(\xi_n)}{h_n} \geq o(1), \tag{5.52}$$

and hence $f^o(x^*; u) \geq 0$. □

Part III

Software Reference

Chapter 6

Using imfil.m

This chapter is about **imfil.m** and its use. As in the earlier chapters, the notation in the code fragments is different from the mathematical notation in the text. So, for example, the initial iterate in the code is x0 instead of x_0, which is the notation we will use in the text.

The full calling sequence for **imfil.m** is

```
[x,histout,complete_history]=
          imfil(x0,f,budget,bounds,options,extra_data);
```

The last two input arguments `options` (§ 6.4) and `extra_data` (§ 6.8) are optional. You must use the `options` argument if you want to use the `extra_data` argument. If the `options` argument is omitted, **imfil.m** will use the defaults for `options` and make `extra_data` an empty array.

The output argument `complete_history` is also optional but is useful for debugging and performance analysis.

6.1   Installation and Testing

Download the MATLAB files for **imfil.m** from

http://www.siam.org/books/se23/

Put **imfil.m** and **imfil_optset.m** in a directory and put that directory in your MATLAB path.

6.2   Input

The input data are

- $x_0 \in R^N$: the initial iterate;
- f : $R^N \to R^M$: the objective function f if $M = 1$ or, in the case of nonlinear least squares problems with $M > 1$, the nonlinear residual F;

- *budget*: the maximum number of function evaluations allowed to the optimization;

- the bounds array `bounds`,

- the `options` structure, and

- the `extra_data` structure.

We will discuss all but the `options` and `extra_data` arguments in this section. We will explain the `options` at length in § 6.4 and how to send extra data to the function in § 6.8.

6.2.1   The Initial Iterate

imfil.m requires a feasible initial iterate. This means that $x0$ must satisfy the bound constraints, i.e.,

$$bounds(j,1) \leq x_0(j) \leq bounds(j,2)$$

for all j, and that $f(x_0)$ must be defined, i.e., f will return a value for $x0$ with $ifail = 0$.

6.2.2   The Input Function f

If the `simple_function` option is off (i.e., $= 0$), the calling sequence for `f` should be

```
[fout,ifail,icount]=f(x);
```

or

```
[fout,ifail,icount]=f(x,h);
```

if your function is **scale-aware** (see § 6.6.3), i.e., can use the scale to manage its own internal control of accuracy. If your function is scale-aware, set the `scale_aware` . option to 1.

You may omit the *ifail* and *icount* arguments if you set the `simple_function` option to 1. In that event, the calling sequence is

```
fout=f(x);
```

or

```
fout=f(x,h).
```

In most cases the procedures for optimization problems and nonlinear least squares problems are the same, so we will express things in terms of f, the objective function for an optimization problem. When the difference between optimization and nonlinear least squares is important, we will discuss both cases.

If $f(x)$ successfully returns a value, *fout* $= f(x)$ should be that value, the failure flag *ifail* should be 0, and *icount* should be an estimate of the cost. **imfil.m** uses *icount* when comparing the cost of the optimization to the **budget** and to build the first column of the **histout** array, and you have the flexibility to assign noninteger values to *icount*. If, for example, a function call fails after performing half of the normal work, you might set *icount* $= .5$.

ifail $= 1$ is the signal that the function cannot return a value, i.e., a **hidden constraint** has been violated. You must return a NaN when this happens. **imfil.m** will eliminate failed points from the stencil when computing the stencil gradient.

If the **parallel** option is on (i.e., set to 1, 'on', or 'yes'), then **imfil.m** will send an array of input arguments to f. For the evaluation of the stencil derivative, **imfil.m** will send the elements of the stencil that do not violate the bound constraints to f before it computes the stencil gradient. During the line search, **imfil.m** will send every point that could be queried in the line search to f all at once, the default being the four points $\{x + \lambda d\}$ for $\lambda = 1, 1/2, 1/4, 1/8$. You can change this by setting the **maxitarm** option (see § 6.11.2).

Your parallel function must be able to accept an $N \times P$ array of P arguments to f and return three $P \times 1$ arrays of values for *fout*, *ifail*, and *icount*. If you are solving a least squares problem where $F : R^N \to R^M$, then *fout* should be $M \times P$. It is your job to construct your function to use what parallelism you have efficiently. The **simple_function** option does the right thing when the **parallel** option is on. If you send f P vectors, then **imfil.m** will set *icount* $= P$.

6.2.3   The Budget

The optimization will terminate when the cumulative cost (as measured by *icost*) exceeds the **budget**. A budget that is too small will force premature termination (as will a list of scales that is too short). A budget that is too large will waste function evaluations and the iteration will make very little progress in the latter stage (see the discussion in § 8.4). The optimization is likely to finish over budget because **imfil.m** does not stop the outer (optimization) loop midstream. The example in Chapter 8 shows how to set the budget and some effects of making the budget (or the number of scales) too small or too large.

6.2.4   The Bounds

The **bounds** array is an $N \times 2$ array with the lower bounds in the first column and the upper bounds in the second column. For example, if $N = 100$ and the bounds are $2 \le x(i) \le 3$, you would use

```
bounds(:,1)=2*ones(100,1); bounds(:,2)=3*ones(100,1);
```

Keep in mind that **imfil.m** requires you to provide finite bounds for all the variables.

6.3   Output and Troubleshooting

The output of **imfil.m** includes x, an approximation of the solution, and two optional ways to look at the history of the iteration. The `histout` array is an optional iteration history. The `complete_history` structure contains every point where **imfil.m** has evaluated f and either the value of f or a failure flag.

6.3.1   The `histout` array

The `histout` array is an $IT \times (N + 5)$-dimensional array, where IT is simply a counter of the number of times the array is updated. The `histout` array is created after the first function evaluation and updated with a new row after each approximate gradient computation.

For optimization problems

$$histout(:, i) = [fcount, fval, \|\nabla f(x, V, h)\|, \|s\|, iarm, x^T]$$

and for nonlinear least squares

$$histout(:, i) = [fcount, F(x)^T F(x)/2, \|DF(x, V, h)^T F(x)\|, \|s\|, iarm, x^T].$$

For each iteration (row) the first five elements are *fcount*, the number of function evaluations so far (the sum of *icount* from each call to f); *fval*, the current value of the objective function; the norm of the stencil gradient; the norm of the step; and `iarm`, the number of times the step length was reduced in the line search for that iteration. The remaining N elements are x^T, where x is the current iteration. We used the `histout` array for the iteration history plots in the book. Keep in mind that the norm of the step is reported in **imfil.m**'s internal scaling (i.e., bounds between 0 and 1).

When **imfil.m** reduces the scale after a stencil failure, **imfil.m** sets $iarm = -1$ in the `histout` to indicate that no quasi-Newton work at all was done.

The example `pid_example_chapter_1.m` in the `Examples/Case_Study_PID` directory of the software collection shows how to use the `histout` array to plot the iteration history.

6.3.2   The `complete_history` Structure

The `complete_history` structure records the successful points (i.e., those for which f returns a value), the values at the successful points, and the points where f failed to return a value. The fields in the structure are `complete_history.good_points`, `complete_history.good_values`, and `complete_history.failed_points`.

imfil.m uses the complete history structure internally to avoid evaluation of f at the same point more than once. This is a possibility if the poll of the points on the stencil is finding better points and the quasi-Newton iteration is not. When the quasi-Newton method succeeds, it is very unlikely that the new point or the stencil around it will have been sampled before.

The example `history_test.m` in the `Examples/Case_Study_PID` directory of the software collection illustrates the use of the `complete_history` structure to examine the difference between the parallel and serial versions of **imfil.m**.

You may disable the `complete_history` structure, and save some storage, by setting the `complete_history` option to 'off' or 0 with

```
options=imfil_optset('complete_history','off',options);
```

Do this only if you are having serious problems with storage. **imfil.m** can be much less efficient with `complete_history` turned off.

6.3.3   Slow Convergence or No Convergence

When the optimization fails to converge or performs poorly, the `histout` array may indicate the reasons. If, for example, you see that $iarm = -1$ for several iterations in a row, the stencil has failed on those iterations. This is an indicator that you could terminate the optimization earlier by either changing `scaledepth` (§ 6.6.1), `target` (§ 6.10.1), `function_delta` (§ 6.10.3), or `stencil_delta` (§ 6.10.2).

If $iarm = maxitarm$ (see § 6.11.2) for several consecutive iterations, then the line search is failing often but the poll is finding better points on the stencil. This is a signal that the quasi-Newton/Gauss–Newton step is poor, and it may be that your function is not well modeled by a smooth surrogate. In that case, **imfil.m** is reverting to a direct search and you may want to reduce `maxitarm`. If this happens only when the scales become small, then the noise in your function may be large enough to render numerical differentiation ineffective. If you can control the accuracy in f, you should do that and make f scale-aware (§ 6.6.3). Your function may also be poorly scaled, and changing `fscale` (§ 6.5.1) can help.

6.4   Setting Options

You can change **imfil.m**'s algorithmic parameters with the `imfil_optset` command. One way to do this is to begin with a call with no arguments:

```
options=imfil_optset;
```

The output of this call is a MATLAB structure with the default options for **imfil.m**. You need do this only once and then use `imfil_optset` to update the options structure you've created. For example, if you want to change `scaledepth` to 20 and use the SR1 quasi-Newton update, you could call `imfil_optset` three times prior to the call to **imfil.m**:

```
options=imfil_optset;
options=imfil_optset('quasi','sr1',options);
options=imfil_optset('scaledepth',20,options);
```

You can also put all the calls to `optset` on a single line when you initialize the `options` structure:

```
options=imfil_optset('quasi','sr1','scaledepth',20);
```

If you wish to use the `options` structure, you add that as an argument to **imfil.m** when you call it. So your call would look like

```
[x,history]=imfil(x0,f,budget,bounds,options);
```

instead of

```
[x,history]=imfil(x0,f,budget,bounds);
```

Many of the options are toggles, which are either on or off. You may turn a toggle on with any of 1, 'on', or 'yes' and off with any of 0, 'no', or 'off'. For example,

```
options=imfil_optset('least_squares',1);
```

and

```
options=imfil_optset('least_squares','yes');
```

are equivalent. Note that 1 is a numerical value and 'yes' is a string.

6.5   The Inner Iteration

The inner iteration is the optimization loop. **imfil.m** solves general bound constrained optimization problems with a quasi-Newton method and nonlinear least squares problems with the Gauss–Newton iteration. You can replace the built-in methods for the inner iteration with the `executive_function` option (see § 7.5).

6.5.1   Scaling f with `fscale`

If the values of $|f|$ are very small or very large, the quality of the difference gradient which **imfil.m** uses in its search can be poor. **imfil.m** attempts to solve this problem by **scaling** the objective function by dividing it by the size of a "typical value," which we call *imfil_fscale*. In order to do this for nonlinear least squares we scale the least squares residual by $\sqrt{fscale}$. We will discuss the optimization case here and present the complete details in § 7.2.1.

The default is

$$imfil\_fscale = 1.2|f(x_0)|,$$

which is usually fine.

If *imfil_fscale* is too large, the inner iteration within **imfil.m** may terminate too soon, and you may fail to exhaust the information in the current scale. This can lead to poor results, or even complete stagnation (i.e., x_0 is never changed).

If *imfil_fscale* is too small, the optimization steps may be too large, and the line search may fail. In this case **imfil.m** becomes a form of coordinate search, and the performance will suffer.

You can change this by setting the `fscale` option. Setting `fscale` to a negative value will tell **imfil.m** to use

$$imfil\_fscale = |fscale||f(x_0)|,$$

so $fscale = -1.2$ is the default. If $fscale > 0$, then

$$imfil\_fscale = fscale.$$

$fscale = 0$ is not a sensible value; if you blunder and set $fscale = 0$, **imfil.m** will restore the default. If $f(x_0) = 0$, then **imfil.m** will set $imfil\_fscale$ to 1.

6.5.2 Quasi-Newton Methods for General Problems

For general optimization problems you may set the `quasi` option to 0 (steepest descent, i.e., the model Hessian is the identity matrix), 'bfgs' (BFGS), or 'sr1' (SR1). The default is $quasi = 'bfgs'$, the BFGS update.

Because **imfil.m** is intended for small problems, **imfil.m** maintains an approximation to the full model Hessian and does not use a sparse or limited-memory [75] formulation of the quasi-Newton methods.

6.5.3 Nonlinear Least Squares

imfil.m will also solve nonlinear least squares problems where the objective function is

$$f(x) = F(x)^T F(x)/2.$$

You tell the code that you have a nonlinear least squares problem by setting `least_squares` option to 1 with the command

```
options=imfil_optset('least_squares',1,options);
```

and write your function so that the **column vector** $F \in R^M$ is returned. **imfil.m** will compute the objective function $F(x)^T F(x)/2$ for you.

The internal nonlinear squares solver in **imfil.m** is a projected damped Gauss–Newton iteration (see § 3.9.2 or [44, 75]).

6.5.4 Which Best Point to Take?

If the current point is x_{base}, the best point in the stencil is x_{min}, and the point selected by the quasi-Newton (or Gauss–Newton) iteration is x_{newt}, **imfil.m** will select x_{newt} to be the new point as long as the line search succeeds, i.e.,

$$f(x_{newt}) < f(x_{base}).$$

If you prefer to let x_{min} be the new point if

$$f(x_{min}) < f(x_{newt}),$$

set `stencil_wins` to 'yes'. The default is 'no'.

This option is useful both for very rough and very smooth problems. If your optimization landscape has severe discontinuities (as does the example in Chapter 9, which was taken from [53]), then setting `stencil_wins` to 'yes' will help you jump over discontinuities. On the other hand, if the objective function is smooth or very nearly so, setting `stencil_wins` to 'yes' will help avoid local minima when the scale h is large and make no difference if h is very small and the quasi-Newton iteration is working well. That is why the `smooth_problems` option (see § 6.6.4) sets `stencil_wins` to 'yes'. The default is 'no' because **imfil.m** is designed to be a hybrid of search and gradient based methods, and setting `stencil_wins` to 'yes' for all scales can obscure the benefits of the quasi-Newton iteration. The reader can try this for the examples in Chapter 8.

6.5.5   Limiting the Quasi-Newton Step

If the quasi-Newton (or Gauss–Newton) step is too long, the line search may fail repeatedly and you will lose the benefits of the quasi-Newton direction. In that case, the iteration will become coordinate search. You may increase the number of step size reductions by changing the `maxitarm` (see § 6.11.2) option from its default of 3, which is a good idea for problems that are very close to smooth problems. Alternatively, the `limit_quasi_newton` option lets you limit the size of the quasi-Newton step before the line search begins. If you set `limit_quasi_newton` to 'yes', the quasi-Newton direction will be no longer than $10h$, where h is the current scale. The default is 'yes', which is a good choice for noisy problems. For nearly smooth problems, 'no' may be better.

6.6   Managing and Using the Scales

6.6.1   Scalestart and Scaledepth

imfil.m samples f on a stencil centered at the current point. The size of that stencil varies at the optimization progresses. The default shape of the stencil is a central difference stencil with $2N$ points. The range of sizes can be controlled by the `scalestart` and `scaledepth` option.

  If the directions in the stencil are vectors $\{v_i\}_{i=1}^m$, **imfil.m** will sample f at the points

$$x_c + h(L_i - U_i)v_i$$

for $1 \le i \le m$. The default vectors are the $2N$ unit vectors in the positive and negative coordinate directions. The **scale** h varies as the optimization progresses. The sequence of scale is

$$\{2^{-n}\}_{n=\texttt{scalestart}}^{\texttt{scaledepth}}.$$

`scaledepth` can be changed with the `imfil_optset` command. The defaults are *scalestart* $= 1$ and *scaledepth* $= 7$. If you see stagnation in the iteration, reducing `scaledepth` will save some effort, but be aware of the risk of early termination.

6.6.2   `custom_scales`

If you want to use a custom sequence of scales $\{h_n\}_{n=1}^{smax}$, you may do so by setting the `custom_scales` array. This is a MATLAB array H with the scales

$$1 > h_1 > h_2 > \cdots > h_{smax} > 0.$$

$h_1 < 1$ is important because **imfil.m** scales the bounds to 0 and 1, so a choice of $h > 1$ would certainly put all points in the stencil outside of the bound constraints. You can be sure that at least one point (other than the center) is within the bounds by setting $h_1 \le 1/2$. You set this option with

```
options=imfil_optset('custom_scales',H,options);
```

6.6.3   Scale-Aware Functions

The `scale_aware` option tells **imfil.m** that your function is scale-aware. This means that f can adjust its internal cost or accuracy with knowledge of the scale h. If `scale_aware` is set to 1, **imfil.m** will use the scale as a second input argument to f. Your function should look like

```
[fout,ifail,icount]=f(x,h);
```

See § 8.3 for an example.

6.6.4   Smooth Problems

If you must apply **imfil.m** to smooth problems, setting `smooth_problem` to 'yes' will adjust several parameters, most importantly the scales. The result is a good, but not optimally tuned, finite difference quasi-Newton (or Gauss–Newton) code. **imfil.m** has been used in this mode to solve suites of artificial test problems [115]. Setting `smooth_problem` to 'yes' is equivalent to this block of MATLAB code:

```
bscales=[.5, .01, .001, .0001, .00001];
options=imfil_optset(...
            'custom_scales',bscales,...
            'stencil_wins','yes',...
            'limit_quasi_newton','no',...
            'armijo_reduction',.25,...
            'maxitarm',5,options);
```

If you use this option, you should probably increase the budget and consider both the default BFGS quasi-Newton method and SR1. Of course, as we said in the introduction, you are better off if you use a code which has been designed for smooth problems.

6.7   Parallel Computing

The **parallel** option tells **imfil.m** that f can be called with multiple arguments and will return a matrix whose columns are the values of f, *ifail*, and *icount*. So if x is an $N \times P$ array of P arguments to f and **parallel** is set to 1, a call to $f(x)$ will return a $1 \times P$ **row** vector of values, a $P \times 1$ vector of cost estimates, and a $P \times 1$ vector of failure flags. You must return a row vector with P columns for consistency with the nonlinear least squares option.

If you are solving a nonlinear least squares problem, where a call to f returns an $M \times 1$ column vector, your parallel function should return an $M \times P$ array of residual values as well as the $P \times 1$ vectors for *iflag* and *icount*. The parallel algorithm is not the same as the serial method because all the line search possibilities are examined at the same time.

The default is $parallel = 0$.

The examples in § 2.5.2 and Chapters 8, 9, and 10 describe ways to make a serial function parallel.

You must keep in mind that this can be much more complicated than simply putting multiple calls to your function inside a parallel for loop (like the MATLAB **parfor** construct). Parallel for loops typically require that the multiple calls to the function do not compete for the same data, and therefore things like global variables inside your function will likely cause the parallel loop to fail.

6.8   Passing Data to f

You can pass data from your calling program directly to f by adding an optional final argument to the call to **imfil.m**. The calling sequence looks like

```
[x,histout,complete_history]=
        imfil(x0,f,budget,bounds,options,extra_data);
```

The MATLAB code and quadrature codes also let you pass data to a function in this way. You must make the extra argument the final argument to f. For example,

```
[fout,ifail,icount]=f(x,extra_data)
```

or, if f is scale-aware,

```
[fout,ifail,icount]=f(x,h,extra_data).
```

Using the optional final argument is a much better idea than communicating with f with global variables. One reason is that global variables can cause problems with parallelism.

In the examples in Chapter 8 the additional argument is a structure which contains parameters for an initial value problem solver and the data for a nonlinear least squares problem. For the example in Chapter 9, we must make the function aware of the file system in order for parallelism to be effective and pass a structure to do that. The example in Chapter 10 is simpler, and only scalars are passed to the function.

6.9   Stencils

imfil.m offers three stencils. You can change from the default centered difference stencil with the `stencil` option. The choices are a one-sided difference stencil, which uses the positive coordinate e_i if $x_c + he_i$ satisfies the bound constraints, and $-e_i$ otherwise, and the **positive basis stencil** [88, 91], which uses the $N+1$ points $\{e_i\}_{i=1}^N$ and

$$v_{N+1} = -\frac{1}{\sqrt{N}} \sum_{i=1}^{N} e_i.$$

The `stencil` options are 0 for the default central-difference stencil, 1 for the one-sided stencil (for compatibility with the old Fortran code), and 2 for the positive basis stencil.

6.9.1   `vstencil`

You may create your own custom stencil by setting the `vstencil` option to a matrix with your directions in the columns.

To do that, create a matrix VS with your directions in the columns, and then

```
options=imfil_optset('vstencil',VS,options).
```

The example `lc_imfil.m` in the `Examples/Linear_Constraints` directory of the software collection shows how to use the `vstencil` option to avoid stagnation when the default stencil directions are insufficient.

6.9.2   `random_stencil`

You can augment the stencil with k random vectors by setting the `random_stencil` option to k. The theory from § 5.6 and [7, 49] will apply if $k \geq 1$.

The default is $k = 0$ (no random vectors) because we have seen better performance overall with the basic centered difference stencil. One reason for this is that more vectors will delay stencil failure and cause the iteration to spend too much time in the line search.

If you suspect that the optimal point is on a constraint boundary, especially a hidden constraint boundary, and are seeing stagnation in the iteration, you might use this option and play with various values of k. Adding as few as one random vector will make the algorithm provably convergent in the sense of § 5.6. This option augments the stencil with k uniformly distributed points on the unit sphere in R^N [96, 104]. See § 7.4.1 for an example of this option's overcoming stagnation on a hidden constraint boundary.

6.10   Terminating the Outer Iteration

Most problems can be solved with the default termination criteria for the optimization (or outer) iteration. However, if the iteration is terminating too soon (i.e., while it's still making progress) or too late (i.e., taking many iterations while making very little progress), there are several things you can do. You may know things that can help **imfil.m** do its job better or may learn things by looking at the iteration history.

The options in this section let you use what you might know about the function to avoid wasted effort (see § 8.4 for an example). Termination is a tricky problem for sampling methods, which is why **imfil.m** offers many—maybe too many—ways to do it.

Two obvious things are changing the list of scales using the `scaledepth` or `custom_scales` options. The `smooth_problems` option, for example, uses the `custom_scales` option to do part of its job. The options we discuss in this section may help you if working on the scales does not, or you have information about your problem that is best communicated to **imfil.m** with these options.

6.10.1 `target`

You may set a `target` value for the optimization. The optimization will terminate once f is below the target. The default value is -10^8, which means that `target` will play no role in the optimization.

6.10.2 `stencil_delta`

If you know how accurate your function is, you may want to terminate once the variation of the function is smaller than your estimate for the error in the function. Setting `stencil_delta` to your estimate of the absolute error in f will terminate the optimization when the maximum absolute difference of function values on the stencil is smaller than `stencil_delta`. To turn this optional termination test on, set the `stencil_delta` option to your estimate of the error. The default is -1, which means the option is off.

6.10.3 `function_delta`

Another way to use your estimate of the function's accuracy is to terminate the outer iteration when the change in best function value from one successful quasi-Newton iteration to the next is small. Setting `function_delta` to a nonzero value will terminate the optimization when the change in best values is less than *function_delta*. This is the approach which helps in the example in § 8.4. The difference between this and the `stencil_delta` options is subtle, but important. If you choose to terminate the iteration when `stencil_delta` is small, you are testing changes in the optimal point. On the other hand, using `function_delta` is testing the quasi-Newton (or Gauss–Newton) iteration by looking at the change in the optimal value. If your problem is nearly smooth, use `function_delta` if the iteration seems to be making very little progress in the terminal phase (i.e., stagnating). On the other hand, if your problem is not smooth (or is a coarse discretization of a smooth problem), `stencil_delta` may be a better choice for solving your stagnation problems. Look at the example in § 8.4 (and experiment with the code) to see how to use these options.

6.10.4 maxfail

The outer iteration will terminate after `maxfail` consecutive line search or stencil failures. The default is 3. You may want to increase this limit if you think the iteration is terminating too soon. On the other hand, if the iteration is making no progress at all in the terminal phase, you may want to decrease `maxfail`, reduce the number of scales, or change the sequence of scales.

6.11 Terminating the Inner Iteration

The nonlinear (or inner) iteration will terminate when the norm of the difference gradient is sufficiently small, when maximum iterations have been taken for the entire iteration, or when stencil failure is detected. All of these termination criteria can be changed, but you should take care before messing about with these options.

6.11.1 maxit

`maxit` is the upper limit on the number of quasi-Newton (or Gauss–Newton) iterations. The default is 50. You may never have to change this limit. Typically the inner iteration will usually terminate with a stencil failure before 50 quasi-Newton iterations. However, for a nearly smooth problem and a small number of scales, 50 might not be enough.

6.11.2 maxitarm

The line search will reduce the step at most `maxitarm` times before returning a failure. The default is 3. The line search is limited in this way for good reason. If your problem is noisy and you don't find something useful after three reductions, you're not likely to do better with more effort. However, if your problem is nearly smooth, you should increase `maxitarm`. The `smooth_problem` option, for example, will increase `maxitarm`.

6.11.3 Noise-Aware Functions and the svarmin Option

A function is **noise-aware** if it can communicate the size of the noise to **imfil.m**. The function does this via an additional output argument. So, the call looks like

```
[fout,ifail,icount,noise_level]=f(x)
```

where `noise_level` is the function's estimate of the noise.

If the `noise_aware` option is 'on', then **imfil.m** uses the estimate of the noise to tighten the criterion for stencil failure. To do this we evaluate the variation of the objective function on the stencil

$$var = \max_j f(x + hv_j) - \min_j f(x + hv_j)$$

and declare stencil failure when

$$var < noise\_level. \tag{6.1}$$

If f is also **scale-aware**, then you may let the `noise_level` depend on the scale h. **imfil.m** is prepared for this and queries the `noise_level` before each inner iteration.

If the noise in f does not depend on h, you may set it directly with the `svarmin` option. If $svarmin > 0$, then the inner iteration will declare stencil failure when $var < svarmin$. This is equivalent to setting `noise_aware` to 1 or 'on' and having f return `svarmin` for `noise_level`.

6.11.4 Terminating the Quasi-Newton Iteration with `termtol`

The quasi-Newton iteration will terminate when

$$\|\nabla f(x, V, h)\| \le \tau h,$$

which is intended to mimic the necessary conditions for optimality. The constant in the termination criteria is scaled with a typical value for f. So

$$\tau = \texttt{imfil\_fscale} * \texttt{termtol}.$$

`imfil_fscale` is a "typical value" of f and is set with the `fscale` option (see § 6.5.1) and plays an important role in **imfil.m**'s internal scaling. `termtol`, on the other hand, affects only the termination of the quasi-Newton loop.

6.12 `verbose`

The `verbose` option lets you watch **imfil.m** at work. If you set $verbose = 1$, you will see that the first five columns of the rows `histout` array appear on the screen as they are computed. The default is $verbose = 0$, which tells **imfil.m** to print only the most serious warnings on the screen.

This is a useful option when troubleshooting, as it is easy to see problems with the line search or stagnation when $verbose = 1$, and then stop the optimization midstream to fix the problems.

Chapter 7

Advanced Options

In this chapter we discuss some options to **imfil.m** which are powerful enough to do harm. Before using some of these options, you must first understand how **imfil.m** manages its scaling and take great care with the calling sequence. Some of these options use **imfil.m**'s internal data structures. You will need to understand how they work to use the advanced options well. This is especially true with the `explore_function` (§ 7.4) and `executive_function` (§ 7.5) options.

7.1 Adding New Directions to the Stencil

You may add new points to the stencil before the computation of the stencil derivative with the `add_new_directions` option. You set this option to the name of the MATLAB function you want **imfil.m** to call before computing the stencil derivative. The `random_stencil` option (see § 6.9.2) is a special case of adding new directions.

If your function is `my_directions.m` and you are updating an existing options structure, you would set the `add_new_directions` option with

```
options=imfil_optset('add_new_directions',@my_directions,options);
```

The calling sequence for your function should be

```
Vnew = my_directions(x, h, V)
```

Vnew is the matrix with the new directions in its columns. In the input, x is the current point, h is the current scale, and V is the current set of directions.

You have to be somewhat careful with this. **imfil.m** will call your function to add directions immediately before computing the stencil derivative and will use these directions in that computation. One use of this option is to capture tangent directions to explicit constraints. It is not the way to do a global search. Use the `explore_function` option (see § 7.4) to make the search more global. **imfil.m** provides x and V in your original coordinates (so do not attempt to scale them yourself). When you return your new directions to **imfil.m** they are rescaled and normalized internally.

The example `lc_imfil_driver.m` in the `Examples/Linear_Constraints` directory of the software collection shows how to use the `add_new_directions` option for a linearly constrained problem. The `tangent_directions` functions in the example follows [92] and uses the tangent directions to a linear constraint when it is nearly active, and ignores the linear constraint otherwise. This avoids stagnation when the stencil directions are insufficient and, unlike using `vstencil`, does not add more directions when they are not needed.

In the example we seek to minimize

$$f(x) = (1/2 - (x)_1)^2 + (1 - (x)_1)^2 (1 - x(2))^2/4$$
$$+ (1/2 - (x)_1)^2 (1 + (x)_2 - 2(x)_2^2)/10, \tag{7.1}$$

subject to the bound constraints

$$0 \le (x)_1 \le 1, \quad 0 \le (x)_2 \le 1$$

and the linear constraint

$$(x)_1 + (x(2) - 1) \ge 1. \tag{7.2}$$

Clearly the minimizer is $x^* = (.5, 1)^T$. With the default options and an initial iterate of $x_0 = (1, 0)^T$ on the constraint boundary, the iteration will stagnate with stencil failure at each iteration.

We incorporate the linear constraint into the objective function with the extreme barrier approach. The code for the objective is

```
function [fout,ifail,icount]=lc_obj(x)
% LC_OBJ
%
% Hardwire the linear constraint x_1 + x_2 >= 1 into
% the objective to apply the extreme barrier approach
% to constraints.
%
if x(1)+x(2) < 1
   fout=NaN;
   ifail=1;
   icount=0;
else
   fout1=(x(1)-.5)^2;
   fout2=.25*(1-x(1))^2*(1 - x(2))^2;
   fout3= .1*(x(1)-.5)^2*(1 + x(2) - 2* x(2)^2);
   fout=fout1+fout2+fout3;
   ifail=0;
   icount=1;
end
```

The terms in the function are designed to put the optimal point at x^* and force stagnation if one uses the standard basis in the positive and negative coordinate directions.

The file `tangent_directions.m` adds the tangent directions to the linear constraint if any point in the stencil violates the constraints.

```
function vnew=tangent_directions(x,h,v)
% TANGENT_DIRECTIONS
%
% This is an example of a way to add new directions.
%
% If any point in the stencil does not satisfy the linear constraints,
% I will add tangent directions to the stencil.
%
% The linear constraints, which we handle with the extreme barrier
% method, are x(1) + x(2)   >= 1
%
vnew=[];
[mv,nv]=size(v);
yesno=1;
for i=1:nv
    x_trial=x + h * v(:,i);
    yesno=yesno*test_constraint(x_trial);
end
if yesno==0
    vnew=zeros(2,2);
%
% The linear constraints are (1, 1)^T x >= 1.
% So a tangent vector is  (-1, 1)^T. We do not have to normalize
% this vector because imfil_core will do that.
%
    vnew(:,1)=[-1,1]';
    vnew(:,2)=-vnew(:,1);
end
yesno


function yesno=test_constraint(x)
%
% No deep thinking here. Either the constraint is violated (yesno = 0) or
% it's not (yesno = 1).
%
val = x(1) + x(2) ;
yesno = (val >= 1);
```

The code `1d_driver.m` compares the use of the `add_new_directions` option to the `vstencil` option (which we discussed in § 4.4.4). To use `vstencil` you would set the options by

```
VS=[0 1; 0 -1; 1 0; -1 0; -1 1; 1 -1]';
```

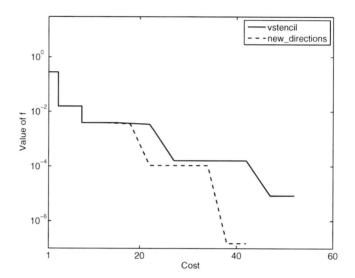

Figure 7.1. *Iteration history with extra directions.*

```
options=imfil_optset('vstencil',VS);
```

and you would use `add_new_directions` in this way:

```
options=imfil_optset('add_new_directions','tangent_directions');
```

Both options will work, but `add_new_directions` is more efficient, as you can see in Figure 7.1. The reason for this is that using the `vstencil` option forces evaluation of f in the extra directions for every iteration, whereas the `add_new_directions` option uses the extra direction only if a point on the stencil violates the constraints.

7.2   The `iteration_data` Structure

imfil.m keeps track of the status of the iteration with the internal `iteration_data` structure. This structure is used heavily within the core of **imfil.m**. The options in § 7.4 and § 7.5 also may need to read some of the fields in that structure. We will explain how these components of `iteration_data` are used in the following sections. For now, we tabulate in Table 7.1 those fields of the structure which are useful for the `executive_function` and `explore_function` options.

Most of the items in Table 7.1 should be clear to a reader who has made it this far in the book. We explain `f_internal` completely in § 7.2.1.

7.2.1   Internal Scaling and `f_internal`

imfil.m scales both the variables and the function value. You will need to understand that scaling to use the `executive_function` and `explore_function` options. This is especially the case if you decide to revert to your original scaling in these

Table 7.1. *Components of the* iteration_data *structure.*

iteration_data.f_internal	Function handle to $f_{internal}$
iteration_data.core_data	Structure passed to $f_{internal}$
iteration_data.complete_history	Complete history structure
iteration_data.xb	xb is the best point found so far in the optimization
iteration_data.funsb	$f_{internal}(xb)$
iteration_data.fobjb	Objective function value at xb
iteration_data.h	Current scale
iteration_data.itc	The inner iteration counter
options	The options structure

functions or rescale the complete_history to examine the iteration in your original coordinates.

Suppose your function is f and your feasible set is

$$\Omega = \{x \in R^N \mid L_i \le (x)_i \le U_i\}.$$

imfil.m begins by transforming Ω to

$$\Omega_{internal} = \{z \in R^N \mid 0 \le (z)_i \le 1\}.$$

The transformation is

$$x = Dz + L,$$

where D is the diagonal matrix with entries

$$D_{ii} = (U_i - L_i).$$

imfil.m also scales f by multiplication by *imfil_fscale* (see § 6.5.1). The inner iteration and the search within **imfil.m** operate on the internal function f_internal. The scaling is

$$f_{internal}(z) = f(Dz + L)/imfil\_fscale$$

for optimization problems and

$$f_{internal}(z) = F(Dz + L)/\sqrt{imfil\_fscale}$$

for nonlinear least squares.

MATLAB functions you write for the options executive_function and explore_function interrupt **imfil.m** midstream and hence will see the internal function and the internal variables. **imfil.m** rescales the history array and the complete_history structure to your original variables after the optimization is complete. If your functions want to see, for example, the complete_history structure in the original coordinates or explore design space in the original coordinates, you will have to do that rescaling yourself. This can get messy and we do not recommend it.

imfil.m informs $f_{internal}$ about the bounds with the core_data structure, which is an extra argument to $f_{internal}$. imfil.m also treats $f_{internal}$ as if it were both scale-aware and noise-aware. The options structure is part of the core_data structure, so the call to $f_{internal}$ will send the correct arguments to f and do the proper things with the other options (such as the parallel option). Your call to $f_{internal}$ should look like

```
[fx,iff,icf,tol]=f_internal(x,h,core_data).
```

You should never have to work with core_data directly, only be prepared to pass it to f_internal. core_data is a substructure of the iteration_data structure, which you get by

```
core_data = iteration_data.core_data;
```

If your original function f is neither scale-aware nor noise-aware, you may set $h = 1$ with no harm and ignore the output argument *tol*.

7.3   Updating the complete_history Structure

The final two advanced options will update the complete_history structure and may want to read it as well. You access this structure as a substructure of the iteration_data structure:

```
complete_history = iteration_data.complete_history
```

You'll need to look at § 6.3.2 if you want to read the data in the structure.

If you wish to write to the structure, you must pass a history structure back to imfil.m. The sections on the executive_function and explore_function options explain where your structure must appear in the list of output arguments. In this section we explain how you must build that structure.

The complete_history structure uses the scaled coordinates, so it is based on evaluations of f_internal. The function values in the structure are scalars for optimization problems and vectors in R^M for nonlinear least squares problems.

We provide a tool build_history in the Imfil_Tools directory which you should use for this purpose. As you evaluate the f_internal you should accumulate the history of the evaluations in three arrays:

- xarray, a matrix with N rows whose columns are the evaluation points;

- funmat, a matrix with M rows whose columns are the evaluations of f_internal at the columns of xvec; and

- failvec, a vector of zeros and ones with failvec(i) = 0 if the evaluation of f_internal succeeded and =1 if the evaluation failed.

Once you have assembled these three arrays, the history of your evaluations can be recorded with

```
my_history = build_history(xarray, funmat, failvec);
```

Your function would send **my_history** back to **imfil.m** as an output argument. **imfil.m** will update the complete_history structure, which you cannot update yourself.

For example, if the parallel option is off and you wish to evaluate f_internal at the P points in xarray, you would build the other arrays by

```
farray=[];
failvec=[];
for i=1:P
    [fout, ifail, icount, tol] = ...
            feval(f_internal, xarray(:,i), h, core_data);
    farray=[farray, fout];
    failvec = [failvec, ifail];
end
```

If the parallel option is on, this is much easier:

```
[farray, failvec, icount, tol] ...
        = feval(f_internal, xarray, h, core_data);
```

Before you build your function's history array, you should determine if the complete_history option is off (rare, but possible). To do this query the **options**, which is substructure of iteration_history:

```
imfil_complete_history = iteration_history.options.complete_history;
if imfil_complete_history == 1
   my_history = build_history(xarray, funmat, failvec);
else
   my_history=[];
end
```

If you want to build your **my_history** array on your own, you must carefully examine the build_history.m file and make sure you construct your history structure correctly. We will invoke a classical warning from [83]: "The world will end if you get this wrong." Therefore, we recommend that you accumulate the function evaluation data and build your **my_history** structure as the last step in your function with the build_history function in the Imfil_Tools directory.

In the sections that follow we show how to use build_history in the context of complete functions, so you can see the context and how to use the iteration_data structure.

7.4 Testing More Points with the explore_function Option

After the inner iteration at a given scale you may explore more globally with the explore_function option. You must write a function to select new points in the scaled feasible set

$$0 \le (x)_i \le 1,$$

evaluate f at these points, and then record the results so that **imfil.m** may update `complete_history` structure. This function is called after the inner iteration and lets you replace the best point from the inner iteration with the results of your exploration or the current best point, whichever is better. You may also want to examine the `complete_history` structure to guide your selection of new points. You may read (but not write to) `complete_history`, which is the

```
iteration_data.complete_history
```

field of the `iteration_data` structure.

The complete history structure is not simple, and you must do the update with care, especially if you rescale back to your original coordinates (which is a very bad idea). The `complete_history` is in the scaled coordinates, so the points all have coordinates in $[0, 1]$. The values of the function are also scaled by *imfil_fscale* (see § 6.5.1). Right before **imfil.m** returns, **imfil.m** rescales the vectors and function values in the `complete_history` structure into the original coordinates, but if you access it before the optimization is complete, you must use the scaled coordinates.

If, for example, your function is `my_search`, you set the option with

```
options=imfil_optset('explore_function',@my_search,options);
```

Your function call must look like

```
[xs, fs, my_cost, explore_history] = ...
     my_search(f_internal,iteration_data,my_search_data);
```

The inputs are `f_internal`, which is **imfil.m**'s internal function, the `iteration_data` structure, and an (optional) final argument `my_search_data` for any data you wish to pass to your explore function. You tell **imfil.m** about `my_search_data` by setting the `explore_data` option:

```
options=imfil_optset('explore_data',my_search_data,options);
```

In the output x_s is the best point from your search, $f_s = f(x_s)$, and *my_cost* is the number of function evaluations your exploration needed. Remember that if you are solving a nonlinear least squares problem, $f_s = F(x_s)$ will be a vector in R^M.

If you wish to turn the `explore_function` option off after you have used it, you may reset the options structure or turn it off explicitly with

```
options=imfil_optset('explore','off');
```

7.4.1   Random Search Example

We return to the example from § 7.1. Our exploration function simply evaluates f at random points in Ω. The number of random points is an extra argument to the function. The MATLAB code for this explore function is in the `Imfil_Tools` directory. The function is more general than we need for this example. In particular, it will do the right things for nonlinear least squares problems and parallel evaluation.

We list the entire function here in order to show you how to get the data you'll need from the `iteration_data` structure, do the main work of the function, and then record the results. The first several lines of the function are devoted to harvesting data. The function evaluation tests for parallelism by querying the `parallel` option and will use parallel evaluation if the `parallel` is on. The next block of code determines if you've found a new best point. Finally, the code builds the `explore_history` structure. You will see this pattern again when we discuss the `executive_function` option in § 7.5.

```
function [xs, fs, my_cost, explore_history] = ...
        random_search(f_internal,iteration_data,my_search_data);
% RANDOM_SEARCH
% function [xs, fs, my_cost, explore_history] = ...
%        random_search(f_internal,iteration_data,my_search_data);
%
% This is an example of an explore_function.
% This function samples f at a few random points and returns the best
% thing it found. I store the number of random points in the
% my_search_data structure.
%
options=iteration_data.options;
parallel = options.parallel;
imfil_complete_history=options.complete_history;
npoints=my_search_data;
xarray=rand(2,npoints);
farray=[];
my_cost=0;
%
% Extract what you need from the structures.
%
% Pass h and core_data to f_internal
%
h=iteration_data.h;
core_data=iteration_data.core_data;
%
% What's the current best point and best objective function value?
%
xb=iteration_data.xb;
funsb=iteration_data.funsb;
fvalb=iteration_data.fobjb;
%
% Am I solving a least squares problem?
%
least_squares=iteration_data.options.least_squares;
%
% Sample the points. Keep the books for the build_history function.
%
```

```
failvec=zeros(1,npoints);
funmat=[];
if parallel == 0
   for i=1:npoints
      x=xarray(:,i);
      [funmati,failvec(i),icount,tol] = ...
                  feval(f_internal,x,h,core_data);
      funmat=[funmat,funmati];
      my_cost=my_cost+icount;
   end
else
   [funmat,fail,icount]=feval(f_internal,xarray,h,core_data);
   my_cost = my_cost+sum(icount);
end
%
% Do the right thing for least squares problems.
%
for i=1:npoints
   if least_squares == 1
      fval=funmat(:,i)'*funmat(:,i)/2;
   else
      fval=funmat(i);
   end
   farray=[farray',fval]';
end


%
% Now see if you've made any progress. If not, return the
% the current best point.
%
[ft,imin]=min(farray);
if failvec(imin) == 0
   xs=xarray(:,imin);
   fs=funmat(:,imin);
else
   fs=funsb;
   xs=xb;
end
%
% Finally, build the explore_data structure.
%
if imfil_complete_history == 1
   explore_history = build_history(xarray, funmat, failvec);
else
   explore_history=[];
end
```

Note that, following the recommendation in § 7.3, explore_function checks to see if the complete_history option is set to the default of yes.

The MATLAB code for the example is

```
Examples/Linear_Constraints/explore_driver.m.
```

In the example we set

```
npoints=10;
options=imfil_optset('explore_function',@my_search,...
                     'explore_data',npoints);
```

which tells the search function to look for the number of points in its last input argument.

Figure 7.2 shows the iteration history with 10 random directions and compares it to one with random_stencil (see § 6.9.2) set to 10. The exploration is faster than the random_stencil option with a similar number of function evaluations, but neither random option is as good as the deterministic methods vstencil (§ 6.9.1) or add_new_directions (§ 7.1), which we compared in Figure 7.1. This is no surprise. The deterministic methods use knowledge of the structure of the problem, and the randomized methods just guess.

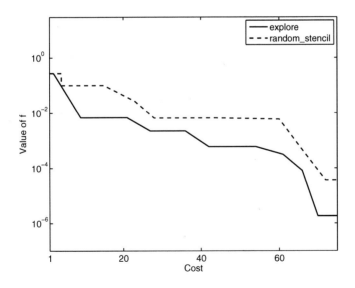

Figure 7.2. *Iteration history with random exploration.*

Finally, in Figure 7.3 we present scatter plots of the successful points from the complete_history structures. As you can see, the random exploration nicely captures the constraint boundary.

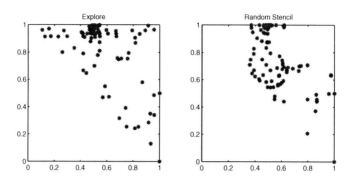

Figure 7.3. *Complete history with random exploration.*

7.5   The Executive Function

If you don't like the gradient-based inner iterations in **imfil.m**, you may replace them with something you like better. The quasi-Newton and Gauss–Newton solvers built into **imfil.m** use the same interface as the one we describe here. These functions take **a single iteration** when called and then return control to **imfil.m**. Your replacement must do that as well.

To do this you use the `executive_function` option to pass **imfil.m** a handle to your solver and, if needed, the `executive_data` option for any data, such as a quasi-Newton model Hessian or a Levenberg–Marquardt parameter, that your solver will update as the iteration progresses. You initialize the data when you set the option. The call to `imfil_optset` will look like

```
options=imfil_optset('executive_function',@my_solver,...
             'executive_data',my_data);
```

where `my_solver` is your solver and `my_data` is your data. `my_data` may be a matrix, a function handle, or any structure you need.

Suppose, for example, your solver is `my_solver.m` and you maintain a quasi-Newton Hessian. If you wish to initialize the Hessian to the identity matrix, you would set up the executive like this:

```
Hess = eye(n,n);
options=imfil_optset('executive_function',@my_solver, ...
                'executive_data',Hess);
```

The data you set with the `executive_data` option may be any matrix or structure you need.

In § 7.5.3 and § 8.4.2 we show how to incorporate the Levenberg–Marquardt method into **imfil.m**. The solver `lev_mar_exec.m` is in the `Imfil_Tools` subdirectory. In the example we initialize the Levenberg–Marquardt parameter to 1 with the call

```
options=imfil_optset('least_squares',1,...
   'executive_function',@lev_mar_exec,'executive_data',1.0);
```

Similarly to the `explore_function` option, your executive function must manage a history structure and conform to a rigid calling sequence. The calling sequence gives you enough information to update a quasi-Newton model Hessian, and you must include that information in the input even if you don't plan to use it. The calling sequence is

```
function [xp, fvalp, funp, fcost, iarm, solver_hist, nfail, new_data] ...
        = my_solver(f, x, fun, sdiff, xc, gc, iteration_data, old_data)
```

7.5.1   Input to the Executive Function

The input string must be somewhat long to accommodate the differing needs of quasi-Newton methods for optimization, which requires some history of the iteration, and other methods, such as Gauss–Newton, which do not. We describe the input arguments in the list below.

- f is a handle to the objective function (optimization) or the least squares residual (nonlinear least squares). **imfil.m** will pass `f_internal` to you with this argument.

- x is the most recent iteration.

- $fun = f(x)$.

- $sdiff$ is the simplex derivative of x. This is the gradient for optimization problems and the Jacobian for nonlinear least squares.

- xc is the previous point at which a simplex derivative was computed.

- gc is the gradient of the objective function at xc.

- `iteration_data` is the iteration data structure (see § 7.2).

- `old_data` is the data your executive function will update.

The input list should be self-explanatory with the exception of the two points and two derivatives, which are for quasi-Newton methods. If your data is a quasi-Newton model Hessian, you will update that model Hessian using x, xc, gc, and $sdiff$ before computing the new point. This is the one time it might help to examine the source of **imfil.m** and, in particular, the function `imfil_qn_update`.

7.5.2   Output from the Executive Function

The output arguments are used by **imfil.m** to update its internal history structures and manage the iteration. The next (and final) list describes the output arguments.

- xp is the new point, which may be the same as x if the iteration fails.

- $fvalp$ is the objective function value at xp. $fvalp = funp$ for optimization problems but not for nonlinear least squares.

- $funp = f(xp)$.

- *fcost* is the total cost of the function evaluations done in your function.

- *iarm* is the counter of step size reductions, Levenberg–Marquardt parameter increments, or other global convergences changes. You may elect to declare a failure when $iarm > maxitarm$.

- `solver_hist` is the update to the `complete_history` structure. You build it the same way as you would the `explore_history` structure. See § 7.3 for the rules.

- *nfail* is the failure flag. $nfail = 0$ if the iteration succeeds. Otherwise $nfail = 1$.

- `new_data` is your function's update of `old_data`. For example, in the quasi-Newton code, `new_data` is the update of the model Hessian. The Gauss–Newton solver does not update the data at all.

An executive function builds its history structure in the same way an explore function does; see § 7.3 and § 7.4.1 for the details.

If you wish to turn the `executive_function` option off after you have used it, you may reset the options structure or turn it off explicitly with

```
options=imfil_optset('executive','off');
```

7.5.3   Levenberg–Marquardt Example

We have put an example in the `Imfil_Tools` directory. The code `lev_mar_exec.m` is am implementation of the Levenberg–Marquardt algorithm. Our implementation follows Algorithm LevMar from § 3.9.4. We apply `lev_mar_exec.m` to an example in § 8.4.2 and will discuss only a few details of the implementation here.

The function definition statement is

```
function [xp, fvalp, funp, fcost, iarm, levmar_hist, nfail, nunew] ...
        = lev_mar_exec(f, x, fun, jac, xc,  gc, ...
            iteration_data, nuold)
```

Here we have followed the directions while naming the history structure and the Levenberg–Marquardt parameter in an appropriate way. This is a least squares computation, so *sdiff* is a stencil Jacobian and we have named the variable accordingly. You should study this code before writing an executive function on your own.

Our implementation is serial. A parallel implementation (which is an exercise for you) would test several candidate Levenberg–Marquardt parameters at one time by evaluating the functions, computing $ared/pred$, and then choosing one parameter or rejecting them all.

One thing any executive function should do is check for input errors. The example function `lev_mar_exec.m` makes sure that the `least_squares` option is on, for instance.

Part IV

Case Studies

Chapter 8

Harmonic Oscillator

The example in this chapter is a **parameter identification** (PID) problem from [9, 75]. In this example $N = 2$. The goal is to identify the damping constant c and spring constant k of a linear spring by minimizing the difference between a numerical prediction and measured data. The experimental scenario is that the spring-mass system will be set into motion by an initial displacement from equilibrium, and measurements of displacements will be taken at equally spaced increments in time.

In this chapter we go into considerable detail about both the application and the MATLAB programming. We do not discuss the programming nearly as much for the other case studies in Chapters 9 and 10 because the application codes were written by others and are best thought of as black-box programs.

The MATLAB codes for this chapter are in the `Examples/Case_Study_PID` subdirectory of the software collection.

8.1 Problem Formulation

We consider an unforced harmonic oscillator where the displacement u is the solution of the initial value problem

$$u'' + cu' + ku = 0; u(0) = u_0, u'(0) = 0, \tag{8.1}$$

on the interval $[0, 10]$. In (8.1) $u' = du/dt$ and $u'' = d^2u/dt^2$.

We let $x = (c, k)^T$ be the vector of unknown parameters and, when the dependence on the parameters needs to be explicit, we will write $u(t : x)$ instead of $u(t)$ for the solution of (8.1). If the displacement is sampled at $\{t_i\}_{i=1}^M$, where $t_i = (i - 1)T/(M - 1)$, and the observations for u are $\{u_i\}_{i=1}^M$, then the objective function is

$$f(x) = \frac{1}{2} \sum_{i=1}^M |u(t_i : x) - u_i|^2. \tag{8.2}$$

We will use MATLAB's `ode15s` [118] to solve (8.1) and use the solution from `ode15s` to compute F. The first step in using `ode15s` is to convert (8.1) to a first-order system for

$$y = \left(\begin{array}{c} u \\ v \end{array} \right) = \left(\begin{array}{c} u \\ u' \end{array} \right).$$

The resulting first-order system is

$$y' = \left(\begin{array}{c} v \\ -cv - ku \end{array} \right), \tag{8.3}$$

with initial data $y(0) = (u_0, 0)^T$.

Here is a MATLAB code for the right side of the differential equation. Note that the parameters c and k are passed to the right side of the differential equation as a third argument. The ode solvers within MATLAB let you pass the parameters to the function in this way by means of an optional final argument in the call to the solvers.

```
function yp=yfunsp(t,y,pid_parms)
%
% simple harmonic oscillator for parameter id example
%
% first-order system form of y'' + c y' + k y = 0
%
yp=zeros(2,1);
yp(1)=y(2);
c=pid_parms(1);
k=pid_parms(2);
yp(2)= - k* y(1) - c*y(2);
```

One might think that it is easier to pass the parameters as MATLAB global variables. However, global variables can cause problems with parallel computing. So, don't do that.

We configure the problem so that the solution is $x^* = (c, k)^T = (1, 1)^T$. For these values of the parameter the solution for $u_0 = 10$ is

$$u = e^{-t/2} \left(10 \cos(\sqrt{3}t) + (5/\sqrt{3}) \sin(\sqrt{3}t) \right).$$

We will begin by letting the data $u_i = u(t_i : x^*)$, where $t_i = (i-1)/100$ for $1 \leq i \leq 101$. So, we compare the output of **ode15s** with the exact solution. The MATLAB codes let you vary the initial data, and the code **yfunex.m** will compute the exact solution for any initial data, c, and k.

The least squares formulation is the most natural. To compute the residual we must solve the initial value problem (8.3) with **ode15s** and then compare the results with the data. In this example we assume that **yfunsp.m** is a file in the MATLAB path. Our main program **driver_pid.m** builds a data structure **pid_info** with several parts:

- **pid_data**: a column vector of size 101 with the data $\{u_i\}_{i=1}^{101}$;

- `time_pts`: the points in time where `ode15s` reports the solution;

- `pid_y0`: the column vector with the initial data, $(10, 0)^T$; and

- `pid_tol`: a scalar for the tolerances for `ode15s`. We set both the relative and absolute tolerances to 10^{-3}.

This structure will not change throughout the optimization. The driver builds the structure with the lines

```
%
% pid_parms contains the zero-residual solution to the noise-free problem
% pid_tol is the tolerance given to ode15s
%
m=100; t0=0; tf=10;
%
% Construct the data for the integration.
% pid_data is a sampling of the "true" solution.
%
pid_parms=[1,1]'; pid_y0=[10,0]'; pid_tol=1.d-3;
time_pts=(0:m)'*(tf-t0)/m+t0;
%
% Find the analytic solution.
%
pid_data=exact_solution(time_pts,pid_y0,pid_parms);
%
% Pack the data into a structure to pass to serial_pidlsq
%
pid_info=struct('pid_y0',pid_y0,'pid_tol',pid_tol,...
                'time_pts',time_pts,'pid_data',pid_data);
```

and then runs **imfil.m** four times to make Figure 8.1. The call to **imfil.m** uses the optional extra argument to send the `pid_data` structure to the function.

```
[x,histout]=imfil(x0,@serial_pidlsq,budget,bounds,options,pid_info);
```

The tolerance for the initial value problem solver is coarse, and we should not expect to be able to reduce the norm of the nonlinear residual by much more than a factor of 10^3, even in the zero-residual case.

The serial code `serial_pidlsq` computes the residual using the `pid_info` structure. This structure is passed to **imfil.m** as an optional final argument, and **imfil.m** sends it directly to `serial_pidlsq` (see § 6.8). The input for `serial_pidlsq` is is the vector of parameters $x = (c, k)^T$ and the `pid_info` structure. `serial_pidlsq` must then pass x to `yfunsp.m`, which it will do with the same optional final argument approach.

```
function [f,ifail,icount]=serial_pidlsq(x,pid_info)
%
% Parameter ID example formulated as nonlinear least squares problem.
%
% Unpack the pid_info structure and get organized.
%
pid_data=pid_info.pid_data;
tol=pid_info.pid_tol;
time_pts=pid_info.time_pts;
y0=pid_info.pid_y0;
%
% Call the integrator only if x is physically reasonable, i.e., if
% x(1) and x(2) are nonnegative. Otherwise, report a failure.
%
ifail=0; icount=1;
if min(x) < 0
   ifail=1; icount=0; f=NaN;
else
   options=odeset('RelTol',tol,'AbsTol',tol,'Jconstant',1);
   [t,y]=ode15s(@yfunsp, time_pts, y0, options, x);
   f=y(:,1)-pid_data(:,1);
end
```

The calling sequence follows the format in § 6.2.2. Note that there is a failure mode. If either c or k is negative, then the spring is not physical and the solution is exponentially increasing. The code traps this and returns without calling the integrator. You could fix this yourself, as we did in the driver program, by making sure that the lower bounds you give to **imfil.m** are all nonnegative.

An integrator like ode15s asks you to provide a local truncation error tolerance via the odeset command. This tolerance controls the accuracy of the integration and thereby the resolution in f. We therefore have an opportunity to experiment with the scale_aware option by letting the accuracy of the integrator depend on the scale. We will do that in § 8.3. For the present we will fix the tolerance to the scalar tol, which serial_pidlsq harvests from the pidinfo structure as pidinfo.pid_tol. We have also used odeset to set the option Jconstant in ode15s to 1, indicating that the differential equation is linear. Finally, we put x in as an optional final argument, which is then passed to yfunsp as its third argument.

If you plan to use the MATLAB initial value problem solvers in your work, study the help files. Typing help odeset and help ode15s at the MATLAB prompt will make this section easier to follow.

8.1.1 Calling imfil.m and Looking at Results

We now show how the simplest call would work. The plots in the upper row of Figure 8.1 reflect a case with an intentionally poor choice of bounds,

$$L = \begin{pmatrix} 2 \\ 0 \end{pmatrix} \quad \text{and} \quad U = \begin{pmatrix} 20 \\ 5 \end{pmatrix},$$

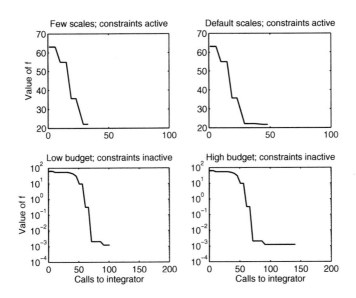

Figure 8.1. *Iteration history: parameter ID.*

which exclude the solution. We gave the optimization a budget of 100 calls to the integrator and an artificially low upper limit of five scales $\{2^{-n}\}_{n=1}^{5}$. We changed the set of scales from the default set $\{2^{-n}\}_{n=1}^{7}$ by using the `imfil_optset` command to change `scaledepth`. The MATLAB commands which follow the construction of the `pid_info` structure are

```
bounds=[2 20; 0 5];
x0=[5,5]'; budget= 100;
options=imfil_optset('scaledepth',5,'least_squares',1);
```

Note that the `least_squares` option is on in this example. Having set the options, the call to **imfil.m** looks like

```
[x,histout]=imfil(x0,@serial_pidlsq,budget,bounds,options,pid_info);
```

Note that the structure `pid_info` is the final argument and is sent directly to `serial_pidlsq`.

As you can see from the plot on the upper left of Figure 8.1, the iteration terminated before the budget had been exhausted. We can return to the default set of scales by reinitializing the `options` structure, but making sure that `least_squares` is still on,

```
options=imfil_optset('least_squares',1);
```

and calling `imfil.m` again. The picture on the upper right reflects the results of this change. Now the optimization requires less than the entire budget, the final value of the objective function is lower (but not by much), and the value of f seems to have stabilized. However, the graph of f also has a flat region earlier in the iteration, but the iteration had not converged at that point. The upper two images in Figure 8.1 illustrate the difficulty in terminating the iteration.

The `histout` array records the progress of the optimization. All the plots were made with the first two columns of the `histout` array. The plots at the top of Figure 8.1 were made with the command

```
plot(histout(:,1),histout(:,2),'-');
```

The first two columns of the `histout` array are the function values and the cumulative cost, measured in this case by calls to `ode15s`. When we look at the plots we see that the optimization has made very little progress after 75 or so calls to `ode15s`. You may modify the example code to add more scales and increase the budget, but the value of the function will decrease only a little, if at all. The reason for this is that we have resolved the optimal point as far as the resolution in the integrator will allow.

The plots on the bottom of Figure 8.1 are from an optimization where the global minimum is within the bounds. Here we set

```
bounds=[0 20; 0 5];
```

The lower left plot in Figure 8.1 shows the progress of the optimization with a budget of 100 and the default set of scales. In this case the budget and the number of scales are sufficient to fully resolve the optimal value. The lower right plot shows the results with a budget of 200 and 20 scales. The results are not very different, and the iteration seems to spend over half the time at the same place. This returns us to the issue of termination. How do we know when to stop the optimization? How can we tell if the budget is too small? Should we change the set of scales? These are open research questions at this time (2011). In § 6.10 and § 6.11 we explored some of the options in **imfil.m** for termination. We will return to the termination issue in this chapter in § 8.4.

8.2   Parallelism

The MATLAB Parallel Toolbox makes parallel computing in MATLAB very accessible, but **NOT EASY**. Even the basic `parfor` loop requires attention to data dependencies and data types. The only way to master the MATLAB tools, or any other parallel environment, is to play with the software, make mistakes, and try to understand the (sometimes opaque) error messages.

8.2.1   Parallelizing the Serial Code

As an example of the use of the `parallel` option, we report on results obtained with the `parfor` loop from the MATLAB Parallel Toolbox. If you don't have that toolbox, you may have to replace the `parfor` loop in `parallel_pidlsq.m` with a `for` loop if you have an older version of MATLAB, and thereby mimic the true parallel version in the sense that you will get the same results as the parallel algorithm *for this example.* However, as a general rule you cannot duplicate the parallel results with this technique. The example in Chapter 10 is one case where you can obtain the (nonreproducible) parallel results only with a truly parallel implementation.

```
function [fa,ifaila,icounta]=parallel_pidlsq(xa,pid_info)
% PARALLEL_PIDLSQ uses parfor to parallelize the serial code.
%
% The code will accept multiple input vectors
% and return a matrix of outputs.
%
[nr,nc]=size(xa);
fa=[];
ifaila=zeros(nc,1);
fcounta=zeros(nc,1);
parfor i=1:nc
    [fap,ifaila(i),icounta(i)]=serial_pidlsq(xa(:,i),pid_info);
    fa=[fa, fap];
end
```

Now one needs to make only a few changes to `driver_pid`. Turn the `parallel` option on and call `parallel_pidlsq.m`. The new lines are

```
options=imfil_optset('parallel',1,options);
[x,histout]=imfil(x0,@parallel_pidlsq,budget,bounds,options);
```

Of course, before using the parallel toolbox, you must create a matlabpool. Here's an example of how one does that:

```
>> matlabpool(8)
Starting matlabpool using the 'local' configuration ...
            connected to 8 labs.
>>
```

In this example, a new matlabpool with eight cores is ready for your parallel job. If you invoke the `matlabpool` command and already have a pool in place, MATLAB will close the existing pool and build a new one.

8.2.2   Looking at the Parallel Results

We will now compare the parallel and serial algorithms. We can do this using the parallel function `parallel_pidlsq` even if the `parallel` option is off. The reason is that if `parallel` is set to 0, the function is evaluated in serial mode. We also compare the least squares formulation to the alternative formulation of using the quasi-Newton optimization algorithm to minimize

$$f(x) = \|F(x)\|^2/2.$$

The MATLAB code for this is `pidobj.m`.

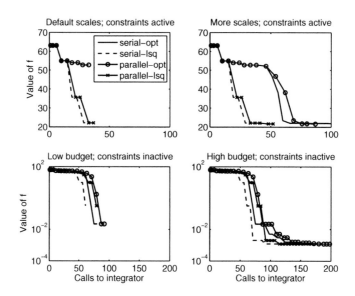

Figure 8.2. *Optimization history: parameter ID revisited.*

```
function [fa,ifaila,icounta]=pidobj(xa,pid_info)
% PIDOBJ calls PARALLEL_PIDLSQ to build an objective
% function that does not use the least squares structure.
%
[nr,nc]=size(xa);
fa=zeros(1,nc);
[fl,ifaila,icounta]=parallel_pidlsq(xa,pid_info);
for i=1:nc
    fa(i)=fl(:,i)'*fl(:,i)/2;
end
```

The call to **imfil.m** would be exactly the same as for the least squares formulation, except you would not turn on the `least_squares` option. Ignoring the least squares structure is a bad idea, as you can see from the plots in Figure 8.2.

We now revisit Figure 8.1 by comparing the serial least squares results in that figure with a serial optimization computation and parallel results. Clearly the least squares formulation is better because of the rapid convergence of the Gauss–Newton iteration for this small-residual problem. As one can see from the lower right plot in Figure 8.2, the number of function evaluations in serial and parallel iteration histories can differ by over 30%, in favor of the serial algorithm, which is not surprising since $N = 2$ and the parallel line search queries three or more possibilities at once. This is an example of the difference between the parallel and serial algorithms. The plots for the active constraint cases show that the parallel and serial algorithms need roughly the same number of iterations. However, the parallel method will, in general, take less time, especially if calls to the function are

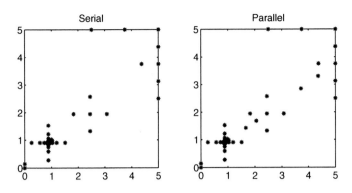

Figure 8.3. *Where is the function evaluated?*

expensive. The code `driver_parallel_pid.m` generates these plots.

One can use the optional output argument `complete_history` (see § 6.3.2) to examine the difference between the parallel and serial algorithms in more detail. The `complete_history` structure records the successful points (i.e., those for which f returns a value), the values at the successful points, and the points where f failed to return a value. The fields in the structure are `complete_history.good_points`, `complete_history.good_values`, and `complete_history.failed_points`. The call to **imfil.m** looks like

```
[x,histout,complete_history]=imfil(x0,f,budget,bounds,options);
```

In Figure 8.3 we plot the good points for both the serial and parallel optimizations for the nonlinear least squares formulation of the parameter identification problem where the constraints are inactive at the solution. This is the computation from the lower right of Figure 8.2. We harvested the data from the `complete_history` structure with the MATLAB program `history_test.m`. This is another view of the example from the lower right corner of Figure 8.2 and shows that the function is evaluated in somewhat different places. The parallel method requires more function evaluations (58) than the serial (51), which is no surprise. Note how the evaluations cluster near the solution in both cases.

8.3 Using the `scale_aware` Option

The `scale_aware` option lets you design functions which can use the scale h to, for example, adjust their internal tolerances. This is very useful if the function is very expensive to evaluate, because then a coarse tolerance early in the iteration can save a significant amount of work. This is an example of how to use **imfil.m** as a multifidelity solver.

Here is a simple example. `sa_serial_pidlsq.m` is a scale-aware version of `serial_pidlsq.m`. The scale-aware version adjusts the tolerance sent to `ode15s` with the formula

$$tol = h^2/10,$$

rather than using the tolerance from the `pid_info` structure. The calling sequence is

```
[f,ifail,icount]=sa_serial_pidlsq(x,h,pid_info)
```

Note that the scale is the second argument to the function. The final argument must always be the extra argument, if you are using one. We have also included a parallel version `sa_parallel_pidlsq.m`.

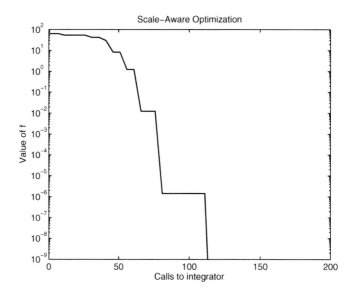

Figure 8.4. *A scale-aware function.*

The driver code `driver_sa.m` repeats the experiment for the unconstrained problem with a budget of 200 and `scaledepth` set to 20. This corresponds to the lower right plot in Figure 8.2. In Figure 8.4 you can see that the value of the function decreases beyond the level at which it stagnated in the earlier computation. The reason for this is that the tolerance for the simulation keeps pace with the decrease in scales.

You should be aware that if you change h for a scale-aware function, you are also changing the function itself. One artifact you may see is an increase in the function value after you change scales. This should not be a surprise, since the function changes each time the tolerance changes, so `imfil` is solving a different problem with every change in scale. This did not happen in this example, but it certainly could have. While you might think that the accuracy of the simulator increases monotonically as the tolerances are tightened, there is no guarantee for that [94].

8.4 Termination Revisited

One striking feature of Figure 8.1 is how the least squares optimizations reached an optimal value by 50 or so iterations, but the iteration did not terminate until the list of scales or the budget was exhausted. Controlling this wasted effort is one of the important and unresolved issues in this field. The examples in this section illustrate some of your options and also show that no single approach will solve all the problems.

If you know something about the size of the noise in your function, **imfil.m** provides three options, `target`, `stencil_delta`, and `function_delta`, which let you exploit that knowledge. For this example, we know the tolerance ($atol = rtol = 10^{-3}$) we've given to **ode15s** and can assume that the function evaluation is not much more accurate than that.

8.4.1 Using `function_delta` to Terminate the Iteration

Setting `function_delta` to $\delta > 0$ with the command

```
options=imfil_optset('function_delta',delta,options);
```

will cause the optimization to terminate as soon as the difference in f between successive iterations is $< \delta$. So, what's δ for this example? In Figure 8.5 we try two values, 10^{-3} and 10^{-6} for the unconstrained case and 5 and 10^{-3} for the constrained (large residual) case. If the constraints are active and the residual at optimality is large, the two values 10^{-3} and 10^{-6} did not save any calls to f, so we tried our luck with 5, a much larger value, which, as you can see from the plot on the left of Figure 8.5, was a very good choice. In the small residual case, setting `function_delta` is also very effective in eliminating the wasted function calls. Note that setting $\delta = 10^{-6}$ results in very little additional progress in this case.

The problem, of course, is picking the appropriate values of the parameter. While this example shows that you can do that, there is no theory to guide you in picking the right parameter, even if you have full knowledge of the errors in f.

Figure 8.5 was generated with the code `driver_term.m`. You might want to modify that code to experiment with the `target` and `stencil_delta` options.

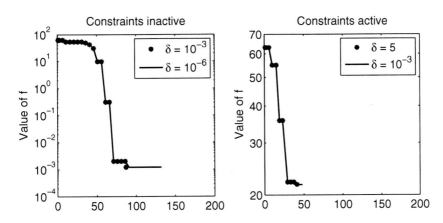

Figure 8.5. *Terminating with* `function_delta`.

8.4.2   Using the Executive Function

In this section we show how the `executive_function` option from § 7.5 can be used to replace the default Gauss–Newton nonlinear least squares solver with a Levenberg–Marquardt code. Our solver `lev_mar_exec.m` is in the `Imfil_Tools` directory. The driver `driver_lm.m` is in the `Examples/Case_Study_PID` directory. Look at § 7.5 and § 7.5.3 to see how the function is defined.

Very little needs to be done to use the solver. The driver compares the Levenberg–Marquardt code to the default Gauss–Newton solver with a scale depth of 20. To use the default solver, set the options with

```
options=imfil_optset('least_squares',1,'scaledepth',20);
```

To use the new solver, you add settings for `executive_function` and `executive_data`. The `executive_data` in this case is the Levenberg–Marquardt parameter, which we initialize to 1. So the options are set with

```
options=imfil_optset('least_squares',1,'scaledepth',20,...
   'executive_function',@lev_mar_exec,'executive_data',1.0);
```

The call to **imfil.m** is the same for either choice

```
[x,histoutgx]=imfilv1(x0,@serial_pidlsq,budget,bounds,options,pid_info);
```

and is the same call we have used throughout this case study.

In Figure 8.6 we compare the two methods. As you can see the Levenberg–Marquardt iteration reduces the residual more rapidly, and significantly so in the case where the constraints are active.

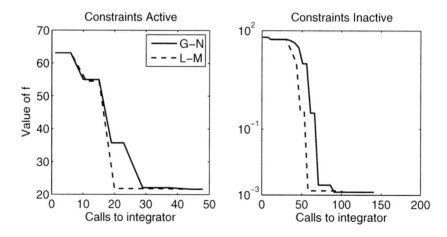

Figure 8.6. *Levenberg–Marquardt executive results.*

Chapter 9

Hydraulic Capture Problem

This case study is one of the examples from [53, 97, 98]. In [97, 98] several "community problems" were posed as test examples from hydrology to evaluate optimization methods. [53] was one of the first papers to compare methods in the context of some of these examples.

It is best that you regard the objective function as a black box. While the MATLAB code for the objective function has been written especially for this book, it uses features of the simulator and properties of the problem in a way that makes the code difficult to modify. The codes for writing, reading, and building the files are excellent examples of this. If you modify the files that call the simulator and create input files for it, you will probably regret it.

The problem in this section is hydraulic capture. The goal is to design an array of wells to capture a contaminant plume at minimum cost. We do that by installing and operating injection and extraction wells to control the direction of the plume migration. We use a gradient control method [3] to define capture in terms of a constraint. The approach is to impose constraints on the pressure differences at an array of points on the plume boundary to drive the contaminants toward an extraction well. The design variables are the locations of the wells and the pumping rates at each well. The formulation in this section manages the number of wells by removing a well from the array if the magnitude of the pumping rate drops below a threshold. Since the cost of installation is significant, the optimization landscape has large discontinuities. There are also other ways to formulate this problem [52, 65, 97].

We used MODFLOW 2000 [64, 100] for the flow simulator. The reader must obtain MODFLOW 2000 from

http://water.usgs.gov/nrp/gwsoftware/modflow2000/modflow2000.html

and install it. The URL also has links to the MODFLOW documentation. You will need a Fortran 90 compiler for the installation and may have to edit a makefile. The results in this chapter were obtained on Apple Macintosh computers running OSX 10.6 and the publicly available `gfortran` Fortran compiler. `gfortran` is available

133

from

http://gcc.gnu.org/fortran/

Everything you need (other than MODFLOW) to reproduce the results in this chapter is in the directory `Examples/Case_Study_HC`.

9.1   Problem Formulation

We take most of this description of the problem formulation from [53, 97]. The physical domain D is an unconfined aquifer. D is a rectangular solid:

$$D = [0, 1000] \times [0, 1000] \times [0, 30]\text{m}.$$

In this section we first describe the equation for saturated flow, which we solve with MODFLOW, and then describe the formulation of the optimization problem.

9.1.1   Saturated Flow Equations

We us MODFLOW to solve the saturated flow equation

$$S_s \frac{\partial h}{\partial t} = \nabla \cdot (K \nabla h) + \mathcal{S} \tag{9.1}$$

for the hydraulic head h, which is a function of $x \in D$ and $t \geq 0$. In (9.1) $S_s = 2.0 \times 10^{-1}$ 1/m is the specific yield, and K is the hydraulic conductivity. In this problem we assume the aquifer is homogeneous (i.e., K is constant) and use

$$K = 5.01 \times 10^{-5}\text{m/s}.$$

The boundary conditions, valid for $t > 0$, are Neumann, no-flow conditions on three faces of D,

$$\left.\frac{\partial h}{\partial x}\right|_{(0,y,z,t)} = \left.\frac{\partial h}{\partial x}\right|_{(x,0,z,t)} = \left.\frac{\partial h}{\partial x}\right|_{(x,y,0,t)} = 0, \tag{9.2}$$

Dirichlet boundary conditions on two faces,

$$h(1000, y, z, t) = 20 - 0.001y \text{ m}, \tag{9.3}$$

$$h(x, 1000, z, t) = 20 - 0.001x \text{ m}, \tag{9.4}$$

and a flux boundary condition (recharge into the aquifer) at the water table where $z = h$,

$$q_z(x, y, z = h, t > 0) = -1.903 \times 10^{-8} \text{ m/s}. \tag{9.5}$$

Here

$$q_z = -K\nabla h$$

is the Darcy flux, and h_s is the steady state solution to the flow problem without wells.

The initial data are

$$S(x, y, z, t = 0) = 0.0 \text{ m}^3/\text{s}, \tag{9.6}$$

$$h(x, y, z, 0) = h_s. \tag{9.7}$$

The ground surface elevation for the unconfined aquifer is $z_{gs} = 30$ m. Here the source term S represents a well model that satisfies

$$\int_D S(t)d\Omega = \sum_{i=1}^{n} Q_i. \tag{9.8}$$

9.1.2 Objective Function

We seek to minimize the cost of installing and operating an array of wells subject to constraints that both reflect operational reality and at the same time should imply capture of a contaminant plume. Given N_W wells our design variables are the location (x_i, y_i) and pumping rate Q_i of well i for $1 \le i \le N_W$. Our formulation will control the number N_W of wells indirectly via the pumping rates.

We seek to minimize the cost of installing and operating an array of N_W wells over a time interval $[0, t_f]$. The cost is a sum

$$f = f^{install} + f^{operate}, \tag{9.9}$$

where

$$f^{install} = \sum_{i=1}^{N_W} c_0 d_i^{b0} + \sum_{i, Q_i < 0.0} c_1 |Q_i^m|^{b_1} (z_{gs} - h^{min})^{b_2} \tag{9.10}$$

and

$$f^{op} = \int_0^{t_f} \left(\sum_{i, Q_i < 0.0} c_2 Q_i (h_i - z_{gs}) + \sum_{i, Q_i > 0.0} c_3 Q_i \right) dt. \tag{9.11}$$

In (9.10) and (9.11) c_i and b_i are model parameters. The depth of well i is $d_i = z_{gs}$. $Q_i^m = 1.5 Q_i$ m^3/s is the design pumping rate. h^{min} is the minimum allowable head, and h_i is the hydraulic head in well i. Injection wells are assumed to operate under gravity feed conditions. In f^c, the first term accounts for drilling and installing all the wells, and the second term is an additional cost for pumps for the extraction wells. In f^o, the term pertaining to the extraction wells includes a lift cost to raise the water to the surface. We tabulate the model parameters in Table 9.1.

The installation cost for a well is roughly $20,000 and the operating cost roughly $1,000. This implies that the optimal design will have as few wells as are needed to meet the constraints. We incorporate this into the optimization by removing a well (i.e., deleting its installation costs from $f^{install}$ and setting the pumping rate to zero) if the absolute value of the pumping rate is sufficiently low. For this application the removal threshold is

$$|Q_i| < 10^{-6} \text{ m}^3/\text{s}. \tag{9.12}$$

Table 9.1. *Model parameters.*

Parameter	Value	Units
c_0	5.5×10^3	$\$/\mathrm{m}^{b_0}$
c_1	5.75×10^3	$\$/[(\mathrm{m}^3/\mathrm{s})^{b_1} \cdot \mathrm{m}^{b_2}]$
c_2	2.90×10^{-4}	$\$/\mathrm{m}^4$
c_3	1.45×10^{-4}	$\$/\mathrm{m}^3$
b_0	0.3	-
b_1	0.45	-
b_2	0.64	-
z_{gs}	30	m
d_i	z_{gs}	m
Q_i^m	$1.5Q_i$	m^3/s

Removing wells in this way introduces large discontinuities into f, as you can see in Figure 9.2.

9.1.3 Constraints

The bounds on the pumping rates are

$$Q^{emax} \leq Q_i \leq Q^{imax}, \quad i = 1, \ldots, N_W, \tag{9.13}$$

where $Q^{emax} < 0$ is the maximum extraction rate and $Q^{imax} > 0$ is the maximum injection rate. We impose a single explicit linear constraint

$$Q_T = \sum_{i=1}^{n} Q_i \geq Q_T^{max}, \tag{9.14}$$

where Q_T^{max} is the maximum allowable total extraction rate.

The bound constraints on the well locations (x_i, y_i) are that the wells be at least $200\,\mathrm{m}$ away from the Dirichlet boundaries, i.e.,

$$0 \leq x_i, y_i \leq 800\,\mathrm{m}. \tag{9.15}$$

The simulator requires that the well locations be at spatial mesh points. Since the optimization may well move these locations to points not on the grid, we round each location to the nearest point on the spatial mesh. This results in a mild discontinuity in the objective function. The major problem with this approach is not the discontinuity but that the rounding may put two wells in the same location, which is not allowed. When this happens we call the point infeasible. We also regard the requirement that two wells not occupy the same location as a hidden constraint.

We also constrain the heads at the wells,

$$h^{min} \leq h_i \leq h^{max}, \quad i = 1, \ldots, N_W, \tag{9.16}$$

where h^{max} is the maximum allowable head, and h^{min} is the minimum allowable head. The rate constraints reflect limits on the pump and well design. The upper bound in (9.16) says that the head must be below the surface elevation and the lower bound limits the drawdown. Note that (9.16) is a nonlinear constraint because the heads are nonlinear functions of the pumping rates and well locations. We regard the head constraints as hidden constraints, because one must run the simulator to test for feasibility.

Table 9.2 lists the values of the many parameters in the constraints on the design variables and the heads.

Table 9.2. *Constraint parameters.*

Parameter	Value	Units
Q_T^{max}	-3.2×10^{-2}	m^3/s
Q^{emax}	-6.4×10^{-3}	m^3/s
Q^{imax}	6.4×10^{-3}	m^3/s
h^{min}	10	m
h^{max}	30	m
d	10^{-4}	m

The final set of constraints are the gradient constraints which are surrogates for hydraulic capture. These say that the pressures are directing the plume inward. The plume was constructed in advance using the USGS MT3DMS [136] code (which you do not need to run the example) by simulating the development of a plume from a point source. The details of the construction of the plume are in [53]. The plume boundary is encoded in the constraints for this problem through the location of the control points. All the data for the plume are included in the data files for the example. You do not need to construct the plume yourself. The output from MT3D is included in the data for the example. We had five control points, as shown in Table 9.3.

Table 9.3. *Control points.*

x (m)	y (m)
180	730
240	770
330	740
390	650
380	540

We defined the 5×10^{-5} kg/m^3 concentration contour line as the boundary of the plume. We formulated the gradient constraints in terms of a finite difference approximation at a set of control points on the boundary of the plume.

Table 9.4. *Initial iterate for hydraulic capture problem.*

x (m)	y (m)	Q (m^3/s)
150.0	750.0	0.0064
400.0	750.0	0.0064
250.0	650.0	-0.0064
250.0	450.0	-0.0064

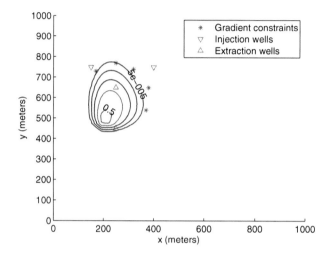

Figure 9.1. *Contaminant plume and wells.*

9.1.4   Design Variables and Initial Iterate

We set $N_W = 4$, which means the number of wells at optimality will be no more than 4 and that the number of design variables is 12. The initial iterate for the well locations and pumping rates (see Table 9.4) is feasible with respect to all constraints.

At the initial iterate, wells 1 and 2 are injecting at the maximum rate and wells 3 and 4 are extracting at the maximum rate.

Figure 9.1 (taken from [53]) shows the position of the contaminant plume, the control points for the gradient constraints, and the wells at the initial iterate.

Figure 9.2 is an image of the optimization landscape near the initial iterate. We varied $x(1)$ and $Q(1)$ between the bounds, leaving all other variables fixed at the initial values. The landscape has two gaps parallel to the Q-axis. These gaps contain one value of x where two wells overlap but are otherwise due to a failure of the gradient constraint because an injection well is crossing the plume. The missing piece in the upper left of the plot is a result of violation of the bound constraints on the heads. The optimization landscape also shows the discontinuity at the $Q(1) = 0$ line, which indicates that well 1 has been removed because of a low pumping rate.

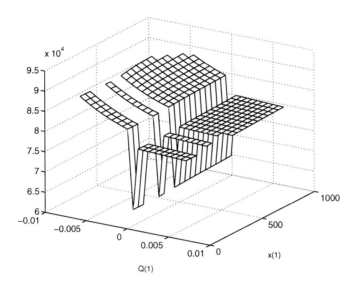

Figure 9.2. *Landscape at initial iterate.*

Implicit filtering can exploit the fact that the initial values of the pumping rates are at the bounds. A consequence of this is that, after scaling, steps of length $h = 1/2$ in a rate will turn off wells. We turn the `stencil_wins` option to let **imfil.m** search for feasible solutions with fewer wells. As we will see in § 9.3, **imfil.m** does this effectively.

9.2 MATLAB Codes and MODFLOW

The MATLAB codes for this case study are in the

`Examples/Case_Study_HC`

directory. We provide objective functions and drivers for both serial and parallel versions of the example. The function evaluation is costly enough to be annoying, and the parallel version is much more pleasant to work with.

9.2.1 Working with MODFLOW

MODFLOW communicates with other programs via file I/O. The advantages of this approach are portability, especially when the files are read and written in plain text, and ease in writing the interface functions. The disadvantage is performance. In the HC problem, MODFLOW writes a file of 10,000 double precision floating point numbers for the pressure head values at the grid points. MODFLOW also requires several data and configuration files.

We manage the files by putting the configuration and data files that you will never change in the

`Examples/Case_Study_HC/DATA`

directory. These files include the grid for the three-dimensional simulation, the data for the initial concentration, the conductivity data, formatting instructions for I/O, and specifications for MODFLOW's internal solvers.

The drivers specify a working directory for the files MODFLOW will need during the optimization. These include a configuration file, which the drivers build for you, the file with the heads that MODFLOW writes, and an input file that the function evaluation must create in response to changes in the positions of the wells and the pumping rates. The default working directory is

```
Examples/Case_Study_HC/hc_tmp
```

The serial function evaluation is `hceval.m` and the parallel function evaluation is `hcevalp.m`. The name of the working directory is an extra argument to the function evaluation, which it needs to share data with MODFLOW. The driver files will create the `hc_tmp` directory for you.

The function evaluation first tests the no-overlap constraint on the well locations and the linear constraint on the pumping rates. If these constraints are satisfied, then the next step is to build the input file for MODFLOW and run MODFLOW. If either of these constraints is not satisfied, then `hceval.m` sets the value to *NaN*, *icost* = 0, and *ifail* = 1.

MODFLOW computes the heads at each grid point and writes that to an output file. The configuration file for MODFLOW contains a formatting directive for the output. We support two formats in the codes in the `Case_Study_HC` directory. In `heval.m` you can set

```
output_method='binary';
```

or

```
output_method='ascii';
```

If you chose "binary," then MODFLOW will use an unformatted Fortran write [26] to create the output file. The advantage of this choice is that the output file is smaller and the reads and writes take less time. The optimization is significantly faster if you use this option. The disadvantage is that the MATLAB code to read the output of MODFLOW may not be portable. The `ascii` option uses a standard Fortran format statement to write the file, and we can use a standard MATLAB format command to read it. This way is slower, but portable. The codes in `Case_Study_HC` are using `ascii`. The optimization gets slightly different results from the two output methods because the precisions differ. The results also differ slightly from those in [53], which used a different version of MODFLOW and different configuration files.

After the call to MODFLOW, `hceval.m` tests the remaining constraints. If `hceval.m` detects infeasibility, then `hceval.m` sets the value to *NaN*, *icost* = 1, and *ifail* = 1. Setting *icost* = 1 means that `hceval.m` had to run MODFLOW to test the constraint. If all constraints are satisfied, `hceval.m` computes the cost and returns.

MODFLOW's files make it harder to manage parallelism. If you are running several copies of MODFLOW at once, the files must have different names. We name the files by adding the index of the `parfor` loop to the file names. If you run the parallel version of the code `hc_driver_parallel.m` and then look inside the `hc_tmp` subdirectory, you will see dozens of input, output, and configuration files. We connect the loop index with the file name by giving `hceval` an optional argument, the process identifier (`pid`), which is used to name the files. That argument is not used in the serial version. In the parallel version we use the optional argument in the `parfor` loop

```
parfor i=1:nv
    [fval(i),ifail(i),icost(i)]=hceval(v(:,i),working_directory,i);
end
```

Here v is the array of nv arguments to `hceval.m`.

We designed the codes in this section for a simple desktop file system. Distributed file systems make it a challenge to synchronize the read and the write, and you may have to take care to make sure that the write is complete before you begin to read the file. See [10] for a discussion of this issue.

9.3   Results

We used the parallel function evaluation and driver

```
Examples/Case_Study_HC/hc_driver_parallel.m
Examples/Case_Study_HC/hcevalp.m
```

to obtain the results in this section. The serial versions

```
Examples/Case_Study_HC/hc_driver.m
Examples/Case_Study_HC/hceval.m
```

take much longer to run. We used MATLAB v 7.11.0.584 (R2010b) on an eight-core Apple MAC Pro. The driver also made the plots in Figures 9.2, 9.3, and 9.4. We made the landscapes with

```
Examples/Case_Study_HC/plot_landscape.m.
```

We used settings very similar to the ones from [53] and obtained very similar results. We use the default options except for `fscale`, `parallel`, and `stencil_wins`. We used $fscale = 9.6 \times 10^4$, the value from [53], and set `parallel` and `stencil_wins` to 1. We allocated a budget of 500 calls to MODFLOW.

After the optimization all the wells except well 3 were shut off. The final result was

$$x(3) = 257.7, \quad y(3) = 634.7, \quad \text{and} \quad Q(3) = -5.35 \times 10^{-4}.$$

The optimization used 360 function evaluations. The optimized cost was \$23,031.

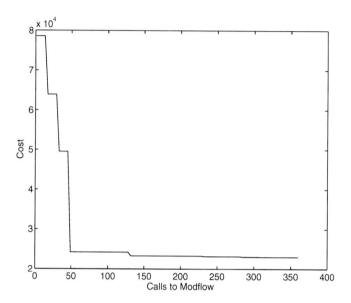

Figure 9.3. *Iteration history.*

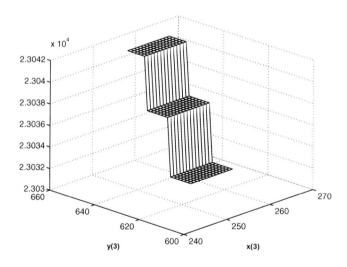

Figure 9.4. *Landscape near converged result.*

The iteration history in Figure 9.3 clearly shows how **imfil.m** identified the discontinuities and did the right thing. This was no accident, as the bounds and initial iterate were designed to let the stencil test for feasible points with wells turned off. However, as more wells are removed, it becomes harder to find a feasible point, as you can see in Figure 9.4.

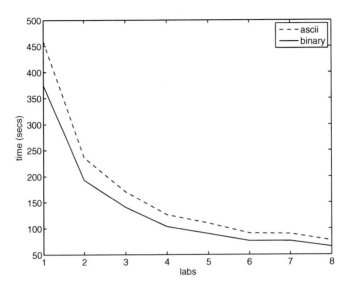

Figure 9.5. *Parallel performance.*

Figure 9.4 is a plot of the landscape near the solution as a function of the location of the only pumping well. Note that the plot is only over a tiny part of the domain, as you can see from the limits on the axes. The feasible region is very small, because the gradient constraint fails if the well is outside the plume. You can also see the discontinuities that arise from the rounding of the well locations to grid points.

We close this section with a discussion of parallel performance. We used up to eight cores (MATLAB "labs") to create the plots in Figure 9.5 and both the ASCII and binary modes of file I/O. We implemented parallelism with the `parfor` loop. The figure indicates good parallel performance for up to six cores and rapidly decreasing parallel efficiency for more cores than that. For example, for the ASCII mode of I/O, the time for one core (460 secs) is 3.6 times that for four cores (126 secs) and 5.0 times that for six cores (91 secs). The figure also illustrates the performance difference for the two modes of I/O.

Chapter 10

Water Resources Policy

In this chapter we consider the design of a water resources policy. The function in the example contains a stochastic simulation, and the problem has hidden constraints. Our description of the problem is taken from [28, 47, 87]. The MATLAB codes for this case study are in the subdirectory `Examples/Case_Study_Water`.

You will need the function `randsample.m` from the MATLAB statistics toolbox.

10.1 Problem Formulation

The problem is to design a reliable water resources policy for a city in the lower Rio Grande valley (LRGV). The context is that cities typically meet demand by maintaining enough reservoir capacity for all conditions short of the most extreme drought; this leads to overcapacity for much of the time. The sequence of papers [27, 28, 47, 61, 86, 87] sought to design a more flexible policy using market-based transfers.

The simulation, as we will see, uses historical streamflow data from this region. The strategy is to combine permanent water rights, leases, and options in a way that minimizes cost subject to risk and variability constraints. We assume that the city gets its water from a reservoir using one of three allocation mechanisms: permanent water rights, spot market leases, and options.

Permanent water rights are a guaranteed fraction of the new water (reservoir inflows, which depend on rainfall) for a given year. Allocations are made at the end of the month. We will assume that the city's volume of permanent rights is constant throughout the year. We denote the annualized cost by p_R. The unit of water is the acre-foot.

Spot market leases are immediate transfers of water. Lease prices are based on the reservoir level at the end of the month. The resulting new supply may be used in any subsequent month. The reservoir level in this model is based on a random sampling of the historical streamflow data and is therefore a random variable. The lease price in month t_i depends on the reservoir level and is a random variable p_L.

In the model in this chapter, the options are European call options, which are purchased before the beginning of the year and can be exercised on a specified call date (t_X, which is May 31 in this model). The option price p_O and exercise price p_X are based on the spot lease prices p_L in the exercise month. The details of the option pricing can be found in [28].

The model period is one year, beginning on December 31 ($t = 0$). The initial data for the model are the reservoir storage R_0 and the amount of water carried over from the previous year N_{r_0}. The number of permanent water rights N_R and options N_O held by the city at the start of the year are two of the six optimization variables. In the following eleven months $\{t_i\}_{i=1}^{11}$ the simulation approximates status of the reservoir by sampling from data sets for monthly reservoir inflow, outflow, and losses, which in turn determine the city's water supply and the spot market price. The city tracks expected supply and demand within the simulation using historical data and the current levels of supply and demand. The ratio of supply to demand is used to decide if the city should purchase leases or exercise options. These decisions depend on four other optimization variables, which we explain below. We compute expected values by averaging the results of several simulations.

10.1.1   Objective Function

The objective function is the expected annual cost

$$
C = N_R p_R + N_O p_O + E[N_X] p_X + E\left[\sum_{t=0}^{11} N_{L_t} p_{L_t}\right]. \tag{10.1}
$$

In (10.1),

N_R = total volume of permanent rights held by the city (ac-ft),

p_R = annualized price of permanent water rights (dollars/ac-ft),

N_O = volume of options purchased at the beginning of the year (ac-ft),

p_O = price of options (dollars/ac-ft),

N_X = volume of exercised options (ac-ft),

p_X = exercise price of options (dollars/ac-ft),

N_{L_t} = volume of leases purchased at the end of each month t (ac-ft), and

p_{L_t} = lease price in month t (dollars/ac-ft).

The number of exercised options N_X, the number of leases the city purchases N_{L_t}, and the price of the leases p_t are random variables because they depend on the status of the reservoir. The purchasing decisions also depend on some of the optimization variables as we explain in § 10.1.3.

10.1.2   Hydrologic Constraints

We impose several hydrologic constraints. The bound constraints (10.2) on the optimization variables N_O and N_R are enforced by **imfil.m**. The other hydrologic constraints (10.3), (10.4), and (10.6) are enforced internally by the simulation and always hold.

We impose lower bound constraints

$$N_O \geq 0 \quad \text{and} \quad N_R \geq 0 \tag{10.2}$$

on two optimization variables. Their upper bounds depend on the particular problem. The city cannot exercise more options than it buys at $t = 0$. Hence

$$N_X \leq N_O. \tag{10.3}$$

The allocations of reservoir inflows to the city cannot exceed the number of permanent rights the city holds.

$$\sum_{t=0}^{11} N_{r_t} \leq N_R, \tag{10.4}$$

where N_{r_t} is the city's allocation from reservoir inflow.

The reservoir level R_t at time t is computed as part of the simulation in terms of some regional hydrologic variables: i_t, the volume of reservoir inflows for month t, l_t, the volume of reservoir losses (e.g., evaporation) for month t, and o_t, the volume of reservoir outflows for month t (which includes spillage).

We compute R_t with a water balance equations

$$R_t = R_{t-1} + i_t - o_t - l_t \tag{10.5}$$

and require

$$R_{min} \leq R_t \leq R_{max}, \tag{10.6}$$

where $R_{min} < R_{max}$ are the limits on the reservoir level.

Not all of the inflow is available for allocation because of the losses. In [28] we defined an instream loss factor l_I and set

$$n_t = (i_t - l_t) \cdot (1 - l_I). \tag{10.7}$$

In the work reported in [28, 47, 87] we set $l_I = .175$, as we did in the computations reported in § 10.3.

The city's fraction of the monthly allocation from the inflow N_{r_t} is

$$N_{r_t} = n_t \left(\frac{N_R}{\overline{N}_R} \right), \tag{10.8}$$

where $\overline{N}_R$ is the total volume of regional water rights.

10.1.3 Purchasing Decisions

The city's decisions on exercising options or purchasing leases on the spot market depends on supply and the predicted demand. We model the supply in two different ways. Before the exercise month t_X for the options we have

$$S_{t+1} = \sum_{i=0}^{t} N_{r_i} + \sum_{i=0}^{t-1} N_{L_i} - \sum_{i=1}^{t} u_i, \quad 0 \leq t \leq t_X - 1. \tag{10.9}$$

In (10.9) S_t is the supply for month t, N_{r_t} the allocation of reservoir inflows to the city in month t, N_{L_t} the volume of leases the city uses in month t, and u_t the city's usage in month t. After the exercise month,

$$S_{t+1} = \sum_{i=0}^{t} N_{r_i} + \sum_{i=0}^{t-1} N_{L_i} - \sum_{i=1}^{t} u_i + N_X, \qquad t_X \leq t \leq 11, \qquad (10.10)$$

where N_X is the volume of exercised options.

The next step in the decision to purchase leases is to compute the expected future supply $S_{E_{t+1}}$ for month $t+1$,

$$S_{E_{t+1}} = S_{t+1} + \sum_{i=t+1}^{11} E[N_{r_i}], \qquad 0 \leq t \leq 10. \qquad (10.11)$$

Here $E[N_{r_i}]$ is the expected value of N_{r_i} in the future month $i > t$. The November $(t = 11)$ inflows are used to compute the available supply for December, but December inflows are assigned to the following year. So December's available supply and expected supply are equal.

The city's demand d_t in month t is determined by random sampling of historical data. We divide the year into two parts, $0 \leq t \leq t_X - 1$ and $t_X \leq t \leq 11$. In part $k = 1, 2$ of the year, the city will purchases leases if the ratio of expected supply to expected demand is below a threshold α_k, i.e.,

$$\frac{S_{E_{t+1}}}{\sum_{i=t+1}^{12} E[d_i]} \leq \alpha_k. \qquad (10.12)$$

α_1 and α_2 are policy parameters and will be optimization variables. The next stage in the purchase decision is to decide how much water to buy. We want to meet the expect demand with a combination of rights, leases, and options. We do this by buying enough leases to make the expected supply larger than β_k times the expected demand period k, where β_1 and β_2 are two more optimization variables. So in any month $t \neq t_X$, we purchase leases until

$$N_{L_t} + S_{E_{t+1}} = \beta_k \sum_{i=t+1}^{12} E[d_i]. \qquad (10.13)$$

In month t_X, which is the start of period 2, we first decide how many options N_X to exercise and then purchase leases to make up any deficit, so we ask

$$N_{L_t} + S_{E_{t+1}} + N_X = \beta_2 \sum_{i=t+1}^{12} E[d_i]. \qquad (10.14)$$

The last step is to determine how many options to exercise before purchasing leases in month t_X. If the exercise price p_X is larger than the lease price p_{L_t}, then the city will only purchase leases and $N_X = 0$ in (10.14). Otherwise we purchase enough options to meet demand and no leases at all, or, if the number of options N_O is not enough, we set $N_X = N_O$ in (10.14) and then

$$N_{L_t} = \beta_2 \sum_{i=t+1}^{12} E[d_i] - N_O - S_{E_{t+1}}.$$

10.1.4 Evaluating the Objective

The initial conditions for the simulation are R_0, the initial reservoir storage level in December, and f_{R_0} the fraction of the total rights that the city holds (carryover water from the previous year). This means that

$$N_{r_0} = f_{R_0} N_R.$$

In the simulations reported in this chapter,

$$R_0 = 1.5 \times 10^6 \text{ac-ft} \quad \text{and} \quad f_{R_0} = .3. \tag{10.15}$$

The optimization variables are

$$x = (N_R, N_O, \alpha_1, \beta_1, \alpha_2, \beta_2)^T \in R^6.$$

The model computes the expected cost (10.1) by averaging N_S simulations. Each simulation is based on a random sample, with replacement, of the historical data. While each simulation takes very little computer time, the time for drawing the landscapes below was significant. The reason for this is that each of the points on the landscape took $N_S = 100$ simulations and the landscape is on a 33×33 mesh.

The bounds are

$$L = \begin{pmatrix} 20000 \\ 0 \\ .7 \\ .7 \\ .7 \\ .7 \end{pmatrix} \le \begin{pmatrix} N_R \\ N_O \\ \alpha_1 \\ \beta_1 \\ \alpha_2 \\ \beta_2 \end{pmatrix} \le \begin{pmatrix} 40000 \\ 10000 \\ 2.2 \\ 2.2 \\ 2.2 \\ 2.2 \end{pmatrix} = U. \tag{10.16}$$

We will also impose the explicit linear constraints

$$\alpha_i \le \beta_i \text{ for } i = 1, 2. \tag{10.17}$$

The number of simulations affects the results, as you can see in the example in this chapter. To illustrate this we plot a landscape of the expected cost for both $N_S = 100$ and $N_S = 500$ simulations. In the landscapes α_i and β_i are fixed for $i = 1, 2$ as

$$\alpha_1 = 1.1, \quad \alpha_2 = 1.3, \quad \beta_1 = .85, \quad \text{and} \quad \beta_2 = 1.10,$$

while N_R and N_O vary between their bounds. We would expect a landscape similar to Figure 1 in [28], and we get it. In Figure 10.1 we draw the landscape using both 100 and 500 simulations. As you can see the landscape with fewer simulations is

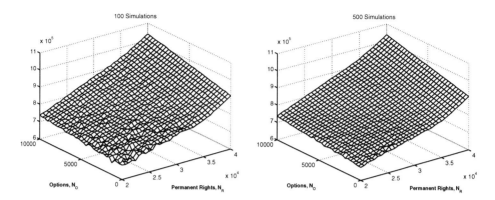

Figure 10.1. *Landscape α_i, β_i fixed.*

rougher than the landscape using five times as many, but not much rougher. So, one might think it is reasonable to use 100 simulations in the optimization. As you will see in § 10.1.5, there are problems with using too few simulations.

The figures in this section and in § 10.1.5 were made with the MATLAB code `LGRV_figs.m`. The landscapes use the MATLAB random number generator. We have reinitialized the random number generator in MATLAB before drawing each landscape. Hence, the landscapes in the figures are reproducible if one executes the simulations on a single processor. Using more processors, with the MATLAB parallel toolbox, for example, will make it impossible to guarantee that the results from two runs of `LGRV_figs.m` are the same. The reason for this is that the model runs for each point on the grid do not occur in a predictable order in a `parfor` loop, so the calls to the MATLAB `rand` function are not the same from run to run. The price for reproducibility is very poor performance, and we advise the reader to use parallelism in MATLAB if possible.

10.1.5   The Reliability and CVaR Constraints

As the cost is computed the simulator computes the expected reliability

$$E[r_f] = 1 - \left(\frac{\text{failures}}{12N_S} \right),$$

where r_f is the monthly reliability and N_S is the number of simulations. A monthly failure means that the city did not meet demand, i.e., $S_t < d_t$. In this chapter we impose the constraint that

$$E[r_f] \geq .995, \tag{10.18}$$

which means one failure every 16.7 years.

The **conditional value at risk** (CVaR) is the mean of the most expensive (top 5%) annual costs in the N_S simulations. We reject a policy that results in a CVaR that is too high and hence impose the constraint

$$\frac{CVaR}{C} \leq 1.1, \tag{10.19}$$

where C is the expected cost (10.1).

In Figures 10.2, 10.3, and 10.4 we illustrate the effect these constraints have on the landscape in Figure 10.1. We will treat these as hidden constraints in § 10.2 and § 10.3 and handle them with the extreme barrier approach. The reader should keep in mind that the reliability and CVaR constraints are stochastic and are not the same when evaluated twice for the same input.

If we insist on a reliability of at least 99.5% we eliminate the possibility of saving cost by having very few permanent rights, and the hidden constraint boundary is the rough edge near the $N_R = 25000$ line.

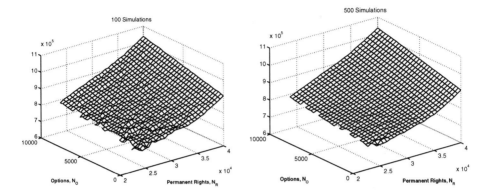

Figure 10.2. *Reliability constraint active.*

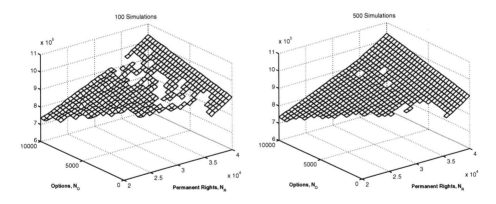

Figure 10.3. *CVaR constraint active.*

To illustrate the variability of the reliability and CVaR constraints and to show how reproducibility of the results is lost when one builds the landscapes in parallel, we draw the landscape from Figure 10.4 using a MATLAB pool of eight labs and a `parfor` loop. The code `LRGV_Parallel.m` manages the parallelism in the same way we did in Chapters 8 and 9. As you can see, the landscape in Figure 10.5 is quite different from the one in Figure 10.4. We will return to this inconvenient

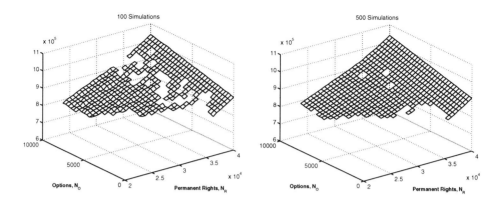

Figure 10.4. *CVaR and reliability constraints active.*

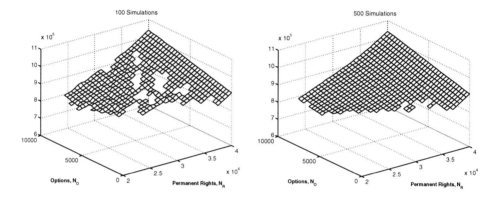

Figure 10.5. *CVaR and reliability constraints active: parallel evaluation.*

truth when we discuss formulation and the solution in § 10.2 and § 10.3.

10.2   Objective Function Code

The code for the objective function LRGV_sim.m was written by Brian Kirsch as part
of the research that led to [28, 47, 86, 87]. The code uses the function randsample.m
from the MATLAB statistics toolbox. You will need randsample.m to run the codes
for this example.

The randomness in the objective has several interesting effects. Clearly the
results of the optimization will be different when run several times with the same
initial iterate. One way to obtain reproducible results is to reinitialize the random
number generator in MATLAB before each optimization, but even this will fail if
the function evaluations are done in parallel, say, by putting a parfor loop around
calls to the serial function evaluation. A more subtle effect is in the constraints. The
reliability and CVaR constraints depend on the internal simulations, which means
a given point may be feasible with one model run and not with the next. imfil.m

handles this possibility with the `complete_history` structure, which means that a point will remain feasible (or infeasible) with respect to a hidden constraint after the first evaluation at that point has been recorded. If one does not do this, the optimal point at one scale may be infeasible when the scale is reduced, and the optimization would fail. This is a subtle issue, and there should be a better way to address it.

10.3   Results

The initial iterate for all the optimizations was

$$
x_0 = \begin{pmatrix} N_R \\ N_O \\ \alpha_1 \\ \beta_1 \\ \alpha_2 \\ \beta_2 \end{pmatrix} = \begin{pmatrix} 40000 \\ 10000 \\ 1.1 \\ 1.3 \\ .85 \\ 1.1 \end{pmatrix},
$$

which is at the high point of the landscapes in the previous sections.

We ran several optimizations, both in serial, where the results are reproducible, and in parallel, where they are not. We allowed the model either $N_S = 100$ or $N_S = 500$ simulations. We allocated a budget of 200 calls to the model. You should keep in mind that the model is not called if the explicit linear constraints (10.17) are violated.

10.3.1   Hiding the Linear Constraints

The results in this section treat the linear constraints as if they were hidden, i.e., with the extreme barrier approach. Our case for doing this is that at optimality the linear constraints are inactive. The tables and figures show that the optimizations converge well and that one can get good results with only 100 simulations, in spite of the ugly landscapes one sees in the figures in § 10.1.5. In Tables 10.1 and 10.2 we show the converged results for serial and parallel optimizations with both 100 and 500 simulations for each model run. The values of the objective function are, from the point of the application, equally good. However, the values of N_O and β_2 are inconsistent. This was also observed in [28, 47].

The parallel optimizations took much less time than the serial ones. It makes little sense to compare the timings directly since the parallel function evaluations are not reproducible. Hence, the parallel optimizations can (and do) give different results when run several times.

Figures 10.6 and 10.7 show the optimization histories for the eight cases. The performance of the optimizers depend only weakly or the number of simulations and the choice of BFGS or SR1.

Table 10.1. *Optimization with BFGS.*

	Serial 100	Serial 500	Parallel 100	Parallel 500
N_R	2.34E+04	2.37E+04	2.51E+04	2.29E+04
N_O	6.44E+03	7.80E+03	4.58E+03	7.73E+03
α_1	1.05E+00	1.10E+00	1.06E+00	1.09E+00
β_1	1.11E+00	1.11E+00	1.11E+00	1.11E+00
α_2	8.50E-01	8.50E-01	8.50E-01	8.50E-01
β_2	1.10E+00	1.48E+00	1.80E+00	1.82E+00
Cost	7.10E+05	7.32E+05	7.10E+05	7.21E+05

Table 10.2. *Optimization with SR1.*

	Serial 100	Serial 500	Parallel 100	Parallel 500
N_R	2.34E+04	2.37E+04	2.34E+04	2.39E+04
N_O	6.44E+03	6.36E+03	8.66E+03	8.29E+03
α_1	1.05E+00	1.10E+00	1.10E+00	1.10E+00
β_1	1.11E+00	1.10E+00	1.10E+00	1.10E+00
α_2	8.50E-01	8.49E-01	8.50E-01	8.56E-01
β_2	1.10E+00	1.82E+00	1.10E+00	1.11E+00
Cost	7.11E+05	7.14E+05	7.35E+05	7.39E+05

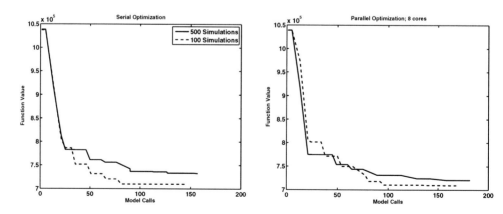

Figure 10.6. *Optimization with BFGS.*

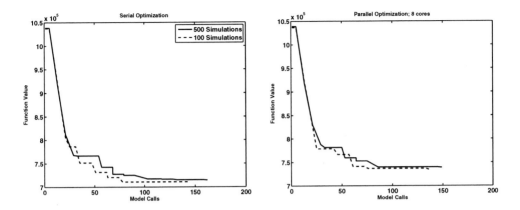

Figure 10.7. *Optimization with SR1.*

10.3.2 Using the `add_new_directions` Option

Using the `add_new_directions` option (7.1) is one way to handle explicit constraints directly. This is easy to do with the simple linear constraints in (10.17), and one option in `LRGV_driver.m` is to do that. In this case, however, the linear constraints are not active at the solution, and using the `add_new_directions` option has almost no effect in most cases, as Tables 10.3 and 10.4 and Figures 10.8 and 10.9 illustrate. In one case, the serial 500 simulation case using BFGS, the constraint $\alpha_1 \leq \beta_1$ is nearly active, and adding the new directions produces an improvement.

Table 10.3. *Optimization with BFGS: more directions.*

	Serial 100	Serial 500	Parallel 100	Parallel 500
N_R	2.57e+04	2.46e+04	2.40e+04	2.32e+04
N_O	5.76e+03	5.50e+03	6.27e+03	8.82e+03
α_1	8.43e-01	1.03e+00	9.73e-01	1.01e+00
β_1	1.07e+00	1.04e+00	9.78e-01	1.19e+00
α_2	7.00e-01	7.45e-01	7.25e-01	7.56e-01
β_2	1.11e+00	1.10e+00	1.05e+00	1.13e+00
Cost	7.20e+05	7.08e+05	7.02e+05	7.43e+05

Table 10.4. *Optimization with SR1: more directions.*

	Serial 100	Serial 500	Parallel 100	Parallel 500
N_R	2.49e+04	2.39e+04	2.37e+04	2.39e+04
N_O	5.11e+03	8.31e+03	6.13e+03	6.43e+03
α_1	1.00e+00	1.03e+00	9.89e-01	1.02e+00
β_1	1.20e+00	1.04e+00	1.08e+00	1.03e+00
α_2	7.55e-01	7.80e-01	7.54e-01	7.67e-01
β_2	1.10e+00	1.10e+00	8.75e-01	1.20e+00
Cost	7.22e+05	7.37e+05	7.03e+05	7.07e+05

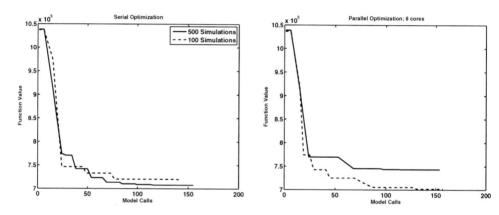

Figure 10.8. *Optimization with BFGS: more directions.*

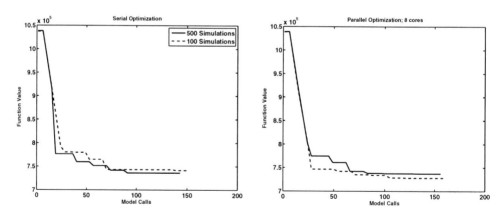

Figure 10.9. *Optimization with SR1: more directions.*

Bibliography

[1] E. AARTS AND J. KORST, *Simulated Annealing and Boltzmann Machines*, John Wiley, New York, 1989.

[2] M. A. ABRAMSON, *Nonlinear Optimization with Mixed Variables and Derivatives-Matlab (NOMADm)*. http://www.gerad.ca/NOMAD/Abramson/ nomadm.html, 2002. Software.

[3] D. P. AHFELD AND A. E. MULLIGAN, *Optimal Design of Flow in Ground-water Systems*, Academic Press, San Diego, 2000.

[4] L. ARMIJO, *Minimization of functions having Lipschitz-continuous first partial derivatives*, Pacific J. Math., 16 (1966), pp. 1–3.

[5] C. AUDET, R. G. CARTER, AND J. E. DENNIS, *Private communication*, 1999.

[6] C. AUDET AND J. E. DENNIS, *Analysis of generalized pattern searches*, SIAM J. Optim., 13 (2003), pp. 889–903.

[7] ——, *Mesh adaptive direct search algorithms for constrained optimization*, SIAM J. Optim., 17 (2006), pp. 188–217.

[8] ——, *A progressive barrier for derivative-free nonlinear programming*, SIAM J. Optim., 20 (2009), pp. 445–472.

[9] H. T. BANKS AND H. T. TRAN, *Mathematical and Experimental Modeling of Physical Processes*, Unpublished lecture notes for Mathematics 573-4, Department of Mathematics, North Carolina State University, 1997.

[10] V. BANNISTER, G. W. HOWELL, C. T. KELLEY, AND E. SILLS, *A case study in using local I/O and GPFS to improve simulation scalability*, in Proceedings of the 8th LCI International Conference on High-Performance Clustered Computing, 2007.

[11] D. P. BERTSEKAS, *On the Goldstein-Levitin-Polyak gradient projection method*, IEEE Trans. Autom. Control, 21 (1976), pp. 174–184.

[12] ——, *Projected Newton methods for optimization problems with simple constraints*, SIAM J. Control Optim., 20 (1982), pp. 221–246.

[13] J. T. BETTS, *Practical Methods for Optimal Control Using Nonlinear Programming*, Advances in Design and Control 3, SIAM, Philadelphia, 2001.

[14] J. T. BETTS, M. J. CARTER, AND W. P. HUFFMAN, *Software for Nonlinear Optimization*, Tech. Rep. MEA-LR-083 R1, Mathematics and Engineering Analysis Library Report, Boeing Information and Support Services, 1997.

[15] Å. BJÖRK, *Numerical Methods for Least Squares Problems*, SIAM, Philadelphia, 1996.

[16] A. J. BOOKER, J. E. DENNIS, P. D. FRANK, D. B. SERAFINI, V. TORCZON, AND M. W. TROSSET, *A rigorous framework for optimization of expensive function by surrogates*, Structural Optim., 17 (1999), pp. 1–13.

[17] D. M. BORTZ AND C. T. KELLEY, *The simplex gradient and noisy optimization problems*, in Computational Methods in Optimal Design and Control, J. T. Borggaard, J. Burns, E. Cliff, and S. Schreck, eds., Progress in Systems and Control Theory 24, Birkhäuser, Boston, 1998, pp. 77–90.

[18] C. G. BROYDEN, *A new double-rank minimization algorithm*, AMS Notices, 16 (1969), p. 670.

[19] M. BUEHREN, *MULTICORE—parallel processing on multiple cores*, MATLAB Central File Exchange, 2007.

[20] R. BYRD, J. C. GILBERT, AND J. NOCEDAL, *A trust region method based on interior point techniques for nonlinear programming*, Math. Program. Ser. A, 89 (2000), pp. 149–185.

[21] R. H. BYRD, H. F. KHALFAN, AND R. B. SCHNABEL, *Analysis of a symmetric rank-one trust region method*, SIAM J. Optim., 6 (1996), pp. 1025–1039.

[22] R. H. BYRD AND J. NOCEDAL, *A tool for the analysis of quasi-Newton methods with application to unconstrained minimization*, SIAM J. Numer. Anal., 26 (1989), pp. 727–739.

[23] R. H. BYRD, J. NOCEDAL, AND Y.-X, YUAN, *Global convergence of a class of quasi-Newton methods on convex problems*, SIAM J. Numer. Anal., 24 (1987), pp. 1171–1190.

[24] R. G. CARTER, *On the global convergence of trust region algorithms using inexact gradient information*, SIAM J. Numer. Anal., 28 (1991), pp. 251–265.

[25] ——, *Numerical experience with a class of algorithms for nonlinear optimization using inexact function and gradient information*, SIAM J. Sci. Comput., 14 (1993), pp. 368–388.

[26] S. J. CHAPMAN, *Fortran 95/2003 for Scientists and Engineers*, McGraw-Hill, Boston, 2008.

[27] G. CHARACKLIS, R. GRIFFIN, AND P. BEDIENT, *Improving the ability of a water market to efficiently manage drought*, Water Resources Research, 35 (1999), pp. 823–832.

[28] G. W. CHARACKLIS, B. R. KIRSCH, J. RAMSEY, K. E. M. DILLARD, AND C. T. KELLEY, *Developing portfolios of water supply transfers*, Water Resources Research, 42 (2006), article W05403.

[29] T. D. CHOI, O. J. ESLINGER, P. GILMORE, A. PATRICK, C. T. KELLEY, AND J. M. GABLONSKY, *IFFCO: Implicit Filtering for Constrained Optimization, Version 2*, Tech. Rep. CRSC-TR99-23, Center for Research in Scientific Computation, North Carolina State University, 1999.

[30] T. D. CHOI AND C. T. KELLEY, *Superlinear convergence and implicit filtering*, SIAM J. Optim., 10 (2000), pp. 1149–1162.

[31] F. H. CLARKE, *Optimization and Nonsmooth Analysis*, Classics in Applied Mathematics 5, SIAM, Philadelphia, 1990.

[32] A. R. CONN, N. I. M. GOULD, AND P. L. TOINT, *Testing a class of methods for solving minimization problems with simple bounds on the variables*, Math. Comp., 50 (1988), pp. 399–430.

[33] ——, *Convergence of quasi-Newton matrices generated by the symmetric rank one update*, Math. Program. A, 50 (1991), pp. 177–195.

[34] ——, *LANCELOT: A Fortran Package for Large-Scale Nonlinear Optimization (Release A)*, Springer Series in Computational Mathematics 17, Springer-Verlag, Heidelberg, Berlin, New York, 1992.

[35] ——, *Trust-Region Methods*, MPS-SIAM Series on Optimization 1, SIAM, Philadelphia, 2000.

[36] A. R. CONN, K. SCHEINBERG, AND P. L. TOINT, *Recent progress in unconstrained optimization without derivatives*, Math. Program. Ser. B, 79 (1997), pp. 397–414.

[37] A. R. CONN, K. SCHEINBERG, AND L. N. VICENTE, *Geometry of sample sets in derivative-free optimization: Polynomial regression and underdetermined interpolation*, IMA J. Numer. Anal., 28 (2008), pp. 721–748.

[38] ——, *Introduction to Derivative-Free Optimization*, MPS-SIAM Series on Optimization, SIAM, Philadelphia, 2009.

[39] A. R. CONN AND P. L. TOINT, *An Algorithm Using Quadratic Interpolation for Unconstrained Derivative-Free Optimization*, Tech. Rep. 95/6, Facultès Universitaires de Namur, 1995.

[40] A. L. CUSTÓDIO AND L. N. VICENTE, *Using sampling and simplex derivatives in pattern search methods*, SIAM J. Optim., 18 (2007), pp. 537–555.

[41] K. DEB, A. PRATAP, S. AGARWAL, AND T. MEYARIVAN, *A fast and elitist multi-objective genetic algorithm: NSGA-I I*, IEEE Transactions on Evolutionary Computation, 6 (2002), pp. 182–197.

[42] J. W. DEMMEL, *Applied Numerical Linear Algebra*, SIAM, Philadelphia, 1997.

[43] J. E. DENNIS, *Nonlinear least squares and equations*, in The State of the Art in Numerical Analysis, D. Jacobs, ed., Academic Press, London, 1977, pp. 269–312.

[44] J. E. DENNIS AND R. B. SCHNABEL, *Numerical Methods for Unconstrained Optimization and Nonlinear Equations*, Classics in Applied Mathematics 16, SIAM, Philadelphia, 1996.

[45] J. E. DENNIS AND V. TORCZON, *Direct search methods on parallel machines*, SIAM J. Optim., 1 (1991), pp. 448–474.

[46] J. E. DENNIS AND H. F. WALKER, *Convergence theorems for least-change secant update methods*, SIAM J. Numer. Anal., 18 (1981), pp. 949–987.

[47] K. E. M. DILLARD, *An Application of Implicit Filtering to Water Resources Management*, Ph.D. thesis, North Carolina State University, Raleigh, 2007.

[48] E. ESKOW AND R. B. SCHNABEL, *Algorithm 695: Software for a new modified Cholesky factorization*, ACM Trans. Math. Softw., 17 (1991), pp. 306–312.

[49] D. E. FINKEL AND C. T. KELLEY, *Convergence analysis of sampling methods for perturbed Lipschitz functions*, Pacific J. Optim., 5 (2009), pp. 339–350.

[50] R. FLETCHER, *A new approach to variable metric methods*, Comput. J., 13 (1970), pp. 317–322.

[51] ——, *Practical Methods of Optimization*, John Wiley, New York, 1987.

[52] K. R. FOWLER, C. T. KELLEY, C. E. KEES, AND C. T. MILLER, *A hydraulic capture application for optimal remediation design*, in Proceedings of Computational Methods in Water Resources XV, C. T. Miller, M. W. Farthing, W. G. Gray, and G. F. Pinter, eds., Elsevier, Amsterdam, 2004, pp. 1149–1158.

[53] K. R. FOWLER, J. P. REESE, C. E. KEES, J. E. DENNIS, C. T. KELLEY, C. T. MILLER, C. AUDET, A. J. BOOKER, G. COUTURE, R. W. DARWIN, M. W. FARTHING, D. E. FINKEL, J. M. GABLONSKY, G. GRAY, AND T. G. KOLDA, *A comparison of derivative-free optimization methods for groundwater supply and hydraulic capture problems*, Adv. Water Resources, 31 (2008), pp. 743–757.

[54] P. E. GILL AND W. MURRAY, *Newton-type methods for unconstrained and linearly constrained optimization*, Math. Program, 28 (1974), pp. 311–350.

[55] P. E. GILL, W. MURRAY, AND M. A. SAUNDERS, *SNOPT: An SQP algorithm for large-scale constrained optimization*, SIAM Rev., 47 (2005), pp. 99–131.

[56] P. E. GILL, W. MURRAY, AND M. H. WRIGHT, *Practical Optimization*, Academic Press, London, 1981.

[57] P. GILMORE, *An Algorithm for Optimizing Functions with Multiple Minima*, Ph.D. thesis, North Carolina State University, Raleigh, North Carolina, 1993.

[58] P. GILMORE AND C. T. KELLEY, *An implicit filtering algorithm for optimization of functions with many local minima*, SIAM J. Optim., 5 (1995), pp. 269–285.

[59] D. GOLDFARB, *A family of variable metric methods derived by variational means*, Math. Comp., 24 (1970), pp. 23–26.

[60] G. H. GOLUB AND C. G. VAN LOAN, *Matrix Computations*, 3rd ed., Johns Hopkins Studies in the Mathematical Sciences, Johns Hopkins University Press, Baltimore, 1996.

[61] R. GRIFFIN AND G. CHARACKLIS, *Issues and trends in Texas water marketing*, Water Resources Update, 121 (2002), pp. 29–33.

[62] M. GU AND S. C. EISENSTAT, *Efficient algorithms for computing a strong rank-revealing QR factorization*, SIAM J. Sci. Comput., 17 (1996), pp. 848–869.

[63] H.-M. GUTMANN, *A radial basis function method for global optimization*, J. Global Optim., 19 (2001), pp. 201–227.

[64] A. W. HARBAUGH, E. R. B. M. C. HILL, AND M. G. MCDONALD, *Modflow-2000, The U.S. Geological Survey Modular Ground-Water Model—User Guide to Modularization Concepts and the Ground-Water Flow Process*, Tech. Rep. Open-File Report 00-92, U.S. Geological Survey, 2000.

[65] T. HEMKER, K. FOWLER, M. FARTHING, AND O. VON STRYK, *A mixed-integer simulation-based optimization approach with surrogate functions in water resources management*, Optim. Engrg., 9 (2008), pp. 341–360.

[66] D. J. HIGHAM, *Trust region algorithms and timestep selection*, SIAM J. Numer. Anal., 37 (1999), pp. 194–210.

[67] J. H. HOLLAND, *Adaption in Natural and Artificial Systems*, University of Michigan Press, Ann Arbor, 1975.

[68] R. HOOKE AND T. A. JEEVES, *'Direct search' solution of numerical and statistical problems*, J. Assoc. Comput. Machinery, 8 (1961), pp. 212–229.

[69] P. D. HOUGH, T. G. KOLDA, AND V. J. TORCZON, *Asynchronous parallel pattern search for nonlinear optimization*, SIAM J. Sci. Comput., 23 (2001), pp. 134–156.

[70] *IEEE Standard for Binary Floating Point Arithmetic*, 754-1885, IEEE, New York, 1985.

[71] I. C. F. IPSEN, *Numerical Matrix Analysis: Linear Systems and Least Squares*, SIAM, Philadelphia, 2009.

[72] I. C. F. IPSEN, C. T. KELLEY, AND S. R. POPE, *Nonlinear least squares problems and subset selection*, SIAM J. Numer. Anal., to appear.

[73] I. T. JOLLIFFE, *Principal Component Analysis*, Springer Series in Statistics, Springer-Verlag, New York, 2002.

[74] C. T. KELLEY, *Iterative Methods for Linear and Nonlinear Equations*, Frontiers in Applied Mathematics 16, SIAM, Philadelphia, 1995.

[75] ———, *Iterative Methods for Optimization*, in Frontiers in Applied Mathematics 18, SIAM, Philadelphia, 1999.

[76] ———, *Implicit filtering and nonlinear least squares problems*, in System Modeling and Optimization XX, E. W. Sachs and R. Tichatschke, eds., Kluwer Academic, Dordrecht, Netherlands, 2003, pp. 71–90.

[77] ———, *Solving Nonlinear Equations with Newton's Method*, Fundamentals of Algorithms 1, SIAM, Philadelphia, 2003.

[78] ———, *Users' Guide for imfil.* http://www4.ncsu.edu/~ctk/MATLAB_IMFIL _v1/manual.pdf, 2011.

[79] C. T. KELLEY, L.-Z. LIAO, L. QI, M. T. CHU, J. P. REESE, AND C. WINTON, *Projected pseudotransient continuation*, SIAM J. Numer. Anal., 46 (2008), pp. 3071–3083.

[80] C. T. KELLEY AND E. W. SACHS, *Local convergence of the symmetric rank-one iteration*, Comput. Optim. Appl., 9 (1998), pp. 43–63.

[81] C. T. KELLEY, E. W. SACHS, AND B. WATSON, *A pointwise quasi-Newton method for unconstrained optimal control problems*, II, J. Optim. Theory Appl., 71 (1991), pp. 535–547.

[82] J. KEPNER, *Parallel MATLAB for Multicore and Mulitnode Computers*, Software Environments and Tools 21, SIAM, Philadelphia, 2009.

[83] B. W. KERNIGHAN AND L. L. CHERRY, *Typesetting Mathematics—User's Guide*, in UNIX Manual, Vol. 2, 7th ed., AT&T Bell Laboratories, Murray Hill, NJ, 1979.

[84] H. F. KHALFAN, R. H. BYRD, AND R. B. SCHNABEL, *A theoretical and experimental study of the symmetric rank-one update*, SIAM J. Optim., 3 (1993), pp. 1–24.

[85] S. KIRKPATRICK, C. D. GEDDAT, AND M. P. VECCHI, *Optimization by simulated annealing*, Science, 220 (1983), pp. 671–680.

[86] B. R. KIRSCH, *Analytical Tools for Integrating Transfers into Water Resource Management Strategies*, Ph.D. thesis, University of North Carolina, Chapel Hill, 2010.

[87] B. R. KIRSCH, G. W. CHARACKLIS, K. E. M. DILLARD, AND C. T. KELLEY, *More efficient optimization of long-term water supply portfolios*, Water Resources Research, 45 (2009), article W03414.

[88] T. G. KOLDA, R. M. LEWIS, AND V. TORCZON, *Optimization by direct search: New perspectives on some classical and modern methods*, SIAM Rev., 45 (2003), pp. 385–482.

[89] S. LE DIGABEL AND C. AUDET, *NOMAD User Guide, Version* 3.1. http://www.gerad.ca/nomad/Project/Home.html, 2009.

[90] K. LEVENBERG, *A method for the solution of certain nonlinear problems in least squares*, Quart. Appl. Math., 4 (1944), pp. 164–168.

[91] R. M. LEWIS AND V. TORCZON, *Rank Ordering and Positive Bases in Pattern Search Algorithms*, Tech. Rep. 96-71, Institute for Computer Applications in Science and Engineering, 1996.

[92] ——, *Pattern search algorithms for linearly constrained minimization*, SIAM J. Optim., 10 (2000), pp. 917–941.

[93] C.-J. LIN AND J. J. MORÉ, *Newton's method for large bound-constrained optimization problems*, SIAM J. Optim., 9 (1999), pp. 1100–1127.

[94] W. LIOEN, J. DE SWART, AND W. VAN DER VEEN, *Test Set for IVP Solvers*, Tech. Rep., Centrum voor Wiskunde en Informatica, Department of Numerical Mathematics, Project Group for Parallel IVP Solvers, 1996.

[95] D. W. MARQUARDT, *An algorithm for least-squares estimation of nonlinear parameters*, J. Soc. Indust. Appl. Math., 11 (1963), pp. 431–441.

[96] G. MARSAGLIA, *Choosing a point from the surface of a sphere*, Ann. Math. Stat., 43 (1972), pp. 645–646.

[97] A. S. MAYER, C. T. KELLEY, AND C. T. MILLER, *Optimal design for problems involving flow and transport phenomena in saturated subsurface systems*, Adv. Water Resources, 12 (2002), pp. 1233–1256.

[98] ———, *Electronic supplement to "Optimal design for problems involving flow and transport phenomena in saturated subsurface systems,"* 2003. http://www.elsevier.com/gej-ng/10/8/34/58/59/41/63/show/index.htt.

[99] D. Q. MAYNE AND E. POLAK, *Nondifferential optimization via adaptive smoothing*, J. Optim. Theory Appl., 43 (1984), pp. 601–613.

[100] M. MCDONALD AND A. HARBAUGH, *A modular three dimensional finite difference groundwater flow model*, U.S. Geological Survey Techniques of Water Resources Investigations, 1988.

[101] C. D. MEYER, *Matrix Analysis and Applied Linear Algebra*, SIAM, Philadelphia, 2000.

[102] R. MIFFLIN, *A superlinearly convergent algorithm for minimization without derivatives*, Math. Program., 9 (1975), pp. 100–117.

[103] J. J. MORÉ, *The Levenberg-Marquardt algorithm: Implementation and theory*, in Numerical Analysis, G. A. Watson, ed., Lecture Notes in Mathematics 630, Springer-Verlag, Berlin, 1977, pp. 105–116.

[104] M. E. MULLER, *A note on a method for generating points uniformly on N-dimensional spheres*, Comm. ACM, 2 (1959), pp. 19–20.

[105] J. A. NELDER AND R. MEAD, *A simplex method for function minimization*, Comput. J., 7 (1965), pp. 308–313.

[106] I. NEWTON, *The Mathematical Papers of Isaac Newton*, D. T. Whiteside, ed., Cambridge University Press, Cambridge, UK, 1967–1976.

[107] R. M. NIXON, *Conversation with John Dean*, 1972.

[108] J. NOCEDAL AND S. J. WRIGHT, *Numerical Optimization*, Springer-Verlag, New York, 1999.

[109] J. M. ORTEGA AND W. C. RHEINBOLDT, *Iterative Solution of Nonlinear Equations in Several Variables*, Academic Press, New York, 1970.

[110] M. L. OVERTON, *Numerical Computing with IEEE Floating Point Arithmetic*, SIAM, Philadelphia, 2001.

[111] K. PEARSON, *On lines and planes of closest fit to systems of points in space*, Philosophical Magazine, 2 (1901).

[112] M. J. D. POWELL, *UOBYQA: Unconstrained optimization by quadratic approximation*, Math. Program., 92 (2002), pp. 555–582.

[113] L. QI AND J. SUN, *A nonsmooth version of Newton's method*, Math. Program., 58 (1993), pp. 353–367.

[114] R. G. REGIS AND C. A. SHOEMAKER, *Local function approximation in evolutionary algorithms for the optimization of costly functions*, IEEE Trans. Evolutionary Comput., 8 (2004), pp. 490–505.

[115] L. M. RIOS AND N. V. SAHINIDIS, *Derivative-Free Optimization: A Review of Algorithms and Comparison of Software Implementations*, unpublished draft manuscript, 2009.

[116] W. RUDIN, *Principles of Mathematical Analysis*, McGraw-Hill, New York, 1953.

[117] R. B. SCHNABEL AND E. ESKOW, *A new modified Cholesky factorization*, SIAM J. Sci. Statist. Comput., 11 (1990), pp. 1136–1158.

[118] L. F. SHAMPINE AND M. W. REICHELT, *The MATLAB ODE suite*, SIAM J. Sci. Comput., 18 (1997), pp. 1–22.

[119] D. F. SHANNO, *Conditioning of quasi-Newton methods for function minimization*, Math. Comp., 24 (1970), pp. 647–657.

[120] M. SRINIVAS AND L. M. PATNAIK, *Genetic algorithms: A survey*, Computer, 27 (1994), pp. 17–27.

[121] G. W. STEWART, *Error and perturbation bounds for subspaces associated with certain eigenvalue problems*, SIAM Rev., 15 (1973), pp. 727–764.

[122] G. W. STEWART, *Introduction to Matrix Computations*, Academic Press, New York, 1973.

[123] D. STONEKING, G. BILBRO, R. TREW, P. GILMORE, AND C. T. KELLEY, *Yield optimization using a GaAs process simulator coupled to a physical device model*, IEEE Trans. Microwave Theory Techniques, 40 (1992), pp. 1353–1363.

[124] D. E. STONEKING, G. L. BILBRO, R. J. TREW, P. GILMORE, AND C. T. KELLEY, *Yield optimization using a GaAs process simulator coupled to a physical device model*, in Proceedings IEEE/Cornell Conference on Advanced Concepts in High Speed Devices and Circuits, 1991, pp. 374–383.

[125] P. L. TOINT, *On large scale nonlinear least squares calculations*, SIAM J. Sci. Statist. Comput., 8 (1987), pp. 416–435.

[126] V. TORCZON, *Multidirectional Search*, Ph.D. thesis, Rice University, Houston, TX, 1989.

[127] L. N. TREFETHEN AND D. BAU, *Numerical Linear Algebra*, SIAM, Philadelphia, 1997.

[128] P. VAN LAARHOVEN AND E. AARTS, *Simulated annealing, theory and practice*, Kluwer Academic, Dordrecht, Netherlands, 1987.

[129] R. J. VANDERBEI, *LOQO: An interior point code for quadratic programming*, Optim. Methods Softw., 11 (1999), pp. 451–484.

[130] B. WATSON, *Quasi-Newton Methoden für Minimierungsprobleme mit strukturierter Hesse-Matrix*, Diploma Thesis, Universität Trier, 1990.

[131] T. A. WINSLOW, R. J. TREW, P. GILMORE, AND C. T. KELLEY, *Doping profiles for optimum class B performance of GaAs mesfet amplifiers*, in Proceedings IEEE/Cornell Conference on Advanced Concepts in High Speed Devices and Circuits, 1991, pp. 188–197.

[132] ——, *Simulated performance optimization of GaAs MESFET amplifiers*, in Proceedings IEEE/Cornell Conference on Advanced Concepts in High Speed Devices and Circuits, 1991, pp. 393–402.

[133] S. J. WRIGHT AND J. N. HOLT, *An inexact Levenbert-Marquardt method for large sparse nonlinear least squares*, J. Austral. Math. Soc. Ser. B, 26 (1985), pp. 387–403.

[134] W. YU, *Positive basis and a class of direct search techniques*, Scientia Sinica, Special Issue on Mathematics, 1 (1979), pp. 53–67.

[135] H. ZHANG, A. R. CONN, AND K. SCHEINBERG, *A derivative-free algorithm for least-squares minimization*, SIAM J. Optim., 20 (2010), pp. 3555–3576.

[136] C. ZHENG AND P. P. WANG, *MT3DMS: A Modular Three-Dimensional Multispecies Transport Model for Simulation of Advection, Dispersion, and Chemical Reactions of Contaminants in Groundwater Systems; Documentation and User's Guide*, 1999.

Index